NEUROSCIENCE AND RELIGION

NEUROSCIENCE AND RELIGION
Brain, Mind, Self, and Soul

Edited by Volney P. Gay

LEXINGTON BOOKS

A DIVISION OF

ROWMAN & LITTLEFIELD PUBLISHERS, INC.

Lanham • Boulder • New York • Toronto • Plymouth, UK

Published by Lexington Books
A division of Rowman & Littlefield Publishers, Inc.
A wholly owned subsidary of The Rowman & Littlefield Publishing Group, Inc.
4501 Forbes Boulevard, Suite 200, Lanham, Maryland 20706
http://www.lexingtonbooks.com

Estover Road, Plymouth PL6 7PY, United Kingdom

British Library Cataloguing in Publication Information Available

Library of Congress Cataloging-in-Publication Data

Neuroscience and religion : brain, mind, self, and soul / edited by Volney P. Gay.
  p. ; cm.
  Includes bibliographical references and index.
  ISBN 978-0-7391-3391-0 (cloth : alk. paper) — ISBN 978-0-7391-3392-7 (pbk. :
alk. paper) — ISBN 978-0-7391-3393-4 (electronic)
  1. Neurosciences—Religious aspects. I. Gay, Volney Patrick [DNLM: 1.
Neuropsychology. 2. Religion and Medicine. 3. Cognition. 4. Ego. 5. Mind-Body
Relations (Metaphysics) 6. Religion and Psychology. WL 103.5 N4949 2009]
  RC343.N446 2009
  616.89—dc22   2009010542

Printed in the United States of America

To Don S. Browning,
the University of Chicago,
for first showing the way

Contents

Figures and Tables

Figures

Tables

Acknowledgments

Funding for the lengthy conversations that went into this volume of essays was provided by the Metanexus Institute and by Vanderbilt University's Academic Venture Capital Fund. We are especially grateful to William Grassie and Eric Weislogel of the Metanexus Institute and to Gordon Gee (former chancellor of Vanderbilt University) and Nicholas Zeppos (former provost and now chancellor of Vanderbilt) for their creative and bold leadership.

1

Neuroscience and Religion: Brain, Mind, Self, and Soul

Volney P. Gay

Science and religion are both still close to their beginnings, with no ends in sight. Science and religion are both destined to grow and change in the millennia that lie ahead of us, perhaps solving some old mysteries, certainly discovering new mysteries of which we yet have no inkling.

—Freeman Dyson[1]

A NUMBER OF AUTHORS have reflected upon the fact that scientists study nature in particular ways, at particular scales, while theologians talk about the Whole or the All. Ian Barbour compares four ways in which one can conceive of scientists and religionists interacting: in Conflict, in Independence, in Dialogue, and a fourth group he labels Integration.[2] Barbour recognizes that adherents of these various approaches confront the problem of locating religion vis-à-vis science. Even religionists and scientists who reject the possibilities of a grand synthesis between science and religion must still give an account of religion and its persistence. A thorough-going reductionist like E. O. Wilson, who labels religion an error to be outgrown by maturing humanity, gives it a place within the evolution of the species. His predecessors, David Hume and Sigmund Freud, also sought to explain how religion emerged and how it should be overcome. Speaking for many atheists, Freud announced that his discipline, psychoanalysis, had only one great enemy, religion, because religious claims pertained to the human psyche, the object upon which psychoanalysis centers. At this psychological level, which Barbour calls the most important religiously, is the human experience of brokenness and loss, redemption and healing.[3]

While some people might feel content with locating religion only at this level (the *meso* level as we describe it below), others are not: "But the centrality of redemption need not lead us to belittle creation, for our personal and social lives are intimately bound to the rest of created order."[4] Our intellectual passions cannot tolerate boundaries upon essential questions: how does it all fit together? This is a metaphysical and a theological question of ancient pedigree. With the rise of the sciences it has evolved: how do scientific truths, derived from all the disciplines, impinge on these truths about human experience?

In the early part of the twenty-first century, we face questions of incommensurate discourses within the sciences and between cultural self-understandings grounded in rival revelations such as those that animate Islam and Christianity. Incommensurability in the sciences typically becomes a problem only for those caught in border disputes; for most scientists, most of the time, the issue is remote from their professional tasks and concerns. However, within the human sciences, and, more so, within religious claims about human being, incommensurability may lie at the center of friction. At their most dangerous, such frictions develop into conflicts that threaten to become religious wars. The rise of economic and demographic globalization is itself a form of enforced unification. With it has come the merging of discourses about essential human values, from the nature of economy and justice to the role of women inside and outside the home and in the new, global, economy. With this unification, we see ideological fractures appear in areas as critical and sensitive as human rights and even the concept of shared humanity.

If we have not yet secured a grand synthesis within physics—nor between physics and chemistry, nor within the biological sciences, nor at the interface between, say, biology and physics, or psychology and economics—it is hardly surprising that we have no synthesis ready to hand with which we can align our diverse religious traditions. And the question of a synthesis between the sciences and religion looks all the more remote. Yet here, as in the other instances, we cannot afford to accept mere agnosticism. Granted that skeptics cannot be proved wrong, our existential interconnectedness and interdependency do not allow us to accept their admonitions to remain silent. Our question, how does it all fit together, is both religious *and* scientific. It emerges in the first moments of reflection: we have a partial grasp of how we fit into time and space, but not a complete grasp. The search for a unified field theory in physics is a contemporary instance of the same urge that drove the first human beings to locate themselves within the boundaries of nature.[5] In our project, we locate many of the sciences and many of the discourses about human being, including religious discourse, rigorously and comprehensively on the basis of size, scope, and meaning. We believe that such an approach can help facilitate a twenty-first-century reconciliation between scientific and religious discourses.[6]

Our practice in the modern university often stands in sharp contrast to the search for unity. Narrowed focus, rigorous exclusion, and relentless self-criticism have produced technical advances not dreamed of by earlier generations. Within information sciences, the pace of advance, miniaturization, and computational power advances inexorably, it seems. But for all the remarkable gains achieved by specialization, there are still far too few chances for discussion among accomplished scholars in the different disciplines. The coming grand synthesis of genomics, protein studies, computer science, and the information sciences is but one example of the beneficial results that collaboration will produce for health care and medicine, not to mention for philosophy and theology.

On the Importance of Scale to Theology and Science: Examples from Neuropsychology, Sociology, and Physics

The nature and size of the objects studied by science and religion matter, because scientific truths (and discoveries) pertain to the scale of those objects. Attempting to generalize to objects below or above that scale proves challenging, if not impossible. The challenge grows exponentially when theologians or metaphysicians attempt to offer valid descriptions of the All, because the All must include everything, from the smallest to the largest. We can clarify this by discussing three concrete instances of the problem of scale within neuropsychology, sociology, and physics.

Neuropsychology

Within contemporary neuropsychology, numerous scientists have labored to map psychological and behavioral events onto a neural substrate: "Cognitive neuroscience is motivated by the precept that a discoverable correspondence exists between mental states and brain states."[7] While some theoreticians may affirm this precept, to bring it into the realm of science one must demonstrate actual instances. To go beyond merely inferring that there is some connection between mental states and brain states, and to discover the precise linkages, neuroscientists must monitor neural events in real time, as they occur during an observed behavior: "A true mapping of neural and mental must be immediate; a mental state can only be supervenient on a neural state in the instant of occurrence."[8]

This is a demanding requirement. It asks the scientist to specify precisely which neural structure (which neurons or sets of neurons) operates during an observed behavior, such as visual attention. Jeff Schall (Ingram Professor of Neuroscience at Vanderbilt University and a participant in the project) describes lucidly how this strict condition leads him and other neuroscientists to conceive of linking propositions. These are propositions that specify how we link observations of the animal's psychological behavior (for example, a monkey looks at a stimulus) to observations of specific neural circuits in the monkey's brain. In their strictest form, linking propositions would state that identical neural states map onto identical cognitive states.[9] On the contra-positive, it must be true that nonidentical neural states map onto nonidentical cognitive states.

The strictest form of mapping holds that every distinct psychological state corresponds exactly to a distinct neural state. If this were so, we should in principle be able to make valid linking statements between any observed mental event and corresponding neural events. However, before affirming so grand a statement, Schall notes that at its current state of development, cognitive neuroscience struggles with issues of scale and size. "At what behavioral and neural scale must the comparison be judged?"[10] It seems very difficult, if not impossible, to specify linking statements at the molecular scale. First, it seems unlikely that the same thought requires us to use the same neuron. Second, given the immense complexity of events at that cellular scale, it would prove difficult to perform the calculations necessary to link events at the level with observed behaviors. It would require us to have a complete map of cellular events that occur in real time. To achieve that remarkable goal we must first create a plausible computational model of cellular events.[11]

If we cannot make linking statements at the molecular level, nor at the level of the single neuron, then we must specify a group of neurons on which we can map mental states. This group of neurons Schall names *the bridge locus,* that set or population of neurons that connect with one another in such ways that their outputs give rise to the observed behavior. Drawing upon a multiyear research project, Schall describes efforts by his research group to map neural events in macaque monkey brains that correspond with saccadic eye movements. (These are fast movements of the eye by which we shift gaze from one object to another.) While much is known about the neural circuits that produce these actions, much is also not known. For example, we don't yet know how the brain terminates saccadic eye movements; theory and evidence tell us that there must be a feedback control circuit, but "fundamental questions remain unanswered."[12] Among the latter are conceptual puzzles as well. For example, how might we map the concepts of "intention" and "deciding" onto neural states? Assuming that we can do so for animals, will this be relevant for humans?[13]

Sociology

Turning to a very different discipline, sociology, we find that issues of scale dominate attempts to link statements about individuals, small groups, and whole societies. Sociologists have developed two different tendencies: one discourages transitions between levels on methodological grounds, and one encourages transitions on theoretical grounds. For over fifty years, a major emphasis in sociology has been to avoid the ecological fallacy.[14] Gary Jensen (chair of sociology and a participant in the project) notes that the ecological fallacy is committed when one uses relationships established using aggregate data representing societies (or ecological territories) as evidence for theories about the behavior of individuals. One dictionary defines it this way: "The bias that may occur because an association observed between variables at an aggregate level does not necessarily represent an association that exists at an individual level; an error of inference due to failure to distinguish between different levels of organization."[15]

Wishing to avoid this fallacy, which is making inferences about "levels of organization" above and below that of a social group, some sociologists generalize neither to individuals nor to nation-states. In sharp contrast, other sociologists, most notably Talcott Parsons, proffer structural accounts of each of the ontological states. Attempting to fulfill Max Weber's vision that sociology would offer a complete theory of human development in its most advanced state, Western capitalism, Parson's theory encompasses each level of human experience.[16]

Because our research group confronts different levels of organization, we need instruction from our sociologist colleagues. It seems unlikely that they alone must struggle with this fallacy. We note, for example, that sociobiologists and their progeny make numerous inferences about human behavior based on the behavior of much lower animals, including insects: "Biologists have shown that a number of apparently altruistic types of behavior can be explained in terms of biological success."[17] This is a strong claim, stuffed with metaphysical assumptions. Before we assent to it, we must first dissect out those assumptions and assess how sociobiologists address the ecological fallacy, among other fallacies. Confronting these and related criticisms, some sociobiologists argue that beyond Darwinian evolution there is a second form, the evolution of *memes*, or what we might call ideas. Aside from the logical and conceptual puzzles this solution presents, we note that it introduces yet another level of entities into the debate. Where mere instincts were, there memes shall be.

The ecological fallacy has loomed so large to sociologists that few have attempted analysis at different levels. Researchers prefer data collected from individuals, because the most common research problem in early sociological research was the use of data from the ecological level to test theories about individual behavior. The fear of committing the ecological fallacy has discouraged scholars from addressing the transitions between macro- and microlevels, and the path taken to deal with the fallacy was to shift to the microlevel.

In response, some contemporary sociologists have proposed that there are patterns or principles that apply at different *fractal* scales, a concept borrowed from the physical sciences and mathematics. For example, Jack Goldstone proposes that social structures are near fractal, showing similar features on macro-micro scales. Goldstone and Useem propose that prison riots are *microrevolutions* and that the theories that explain macrorevolutions can be applied to them. Jensen addresses similar issues that complicate the analysis of relationships over time.[18]

While understandable, the retreat from generalizing between macro- and microstudies means that questions we have about individuals, using sociologically sophisticated data and analyses, will remain unanswered. For example, in his famous study of the sociology of suicide, Émile Durkheim seemed to show that suicide rates were higher in Protestant countries of Europe than in Catholic countries.[19] Assuming that this sociological observation is correct, what can we now say about individual Protestant and Catholic persons in these countries? Durkheim's study spawned many thousands of books and articles attempting to link this sociological fact (again, assuming it merits that title) with theological studies, individual psychology, and epidemiological studies. Each of these must deal with the ecological fallacy, and thus each deals with the issue of scale and hierarchy.

Physics

The history of twentieth-century physics reflects an enormous effort to find unifying theories that span the entire range of physical phenomena, from subatomic particles to the large-scale structures observed by astrophysicists and the universe itself. These theories reflect the essential tension between two fundamentally distinct approaches to experiments and theory: reductionism and emergence. The reductionist view sees physical reality as an onion, with successive layers to be peeled back until one reaches the ultimate constituents of matter. Emergent thinking, on the other hand, reflects the fact that each length scale seems to be describable by entirely self-contained constructs, as if, for example, the rules for describing molecules were quite indifferent to the rules that govern atomic nuclei.

For three-quarters of a century, we have known that Einstein's account of cosmic space-time—and the theory of quantum systems developed by Bohr, Heisenberg, and Schrödinger—are shaped ineluctably by the scales of time and space. The familiar Newtonian world in which we live constitutes a special case described by both theories at the appropriate scales. No physical theory spans the vast conceptual distance from quarks to the cosmos. In part, there are practical difficulties in developing such a theory, because the mathematical framework, the language of each theory, is derived to a significant degree from the constraints imposed by scale. The difficulties are compounded by the realization that the old Laplacian dream of deducing the dynamics of the entire universe from the motion of its constituent parts is frustrated by the sheer complexity of mesoscale and macroscale assemblies of atoms and molecules that go together to make, for example, the crystalline silicon that is the heart of the modern microcomputer. The reality of each length and time scale seems to emerge without reference to the levels above or below.

In some situations, completely different underlying interactions give rise to similar behavior. It can be difficult, if not impossible, to construct a theory that allows the underlying interaction to manifest itself. One cause of this behavior is described by the *central limit theorem* that leads to the prevalence of measurements with normally distributed errors. Repeated measurements of properties of systems in which the central limit theorem is applicable have the same description, though the underlying forces and structure may be as different in nature as the quincunx and the motion of molecules in a gas.

A related difficulty arises from the observation that the scale of a problem may not be well-known at the onset of the study. Thus, it is not always clear whether a concept derived in one context is applicable to another system or whether a new framework must be developed. An example of this arises in the new field of relativistic heavy ions, the goal of which is to produce in the laboratory a state of matter that existed during the first few microseconds after the beginning of the universe. One of the essential conditions is the existence of a very high temperature, billions of times greater than that of the sun. Temperature is a thermodynamic property that requires a statistical description of the entities that make up the system. If the matter produced in these experiments does not reach equilibrium, such a description is invalid. The use of temperature to characterize the system is misleading, and the results of the experiment cannot be described in a way that links old discoveries and new.

The tension between reductionism and emergence can be illustrated in a number of ways. On the complexity frontier, it shows up in the challenge of adapting quantum physics to nanoscale material systems comprising anywhere from a few to a few thousand atoms. In cluster physics, where one deals with a few up to perhaps a hundred atoms, adaptations of atomic physics work reasonably well to describe the properties of the cluster. The reductionist picture that builds upward from atomic properties can, for example, explain how a cluster of gold atoms two nanometers in diameter can be extremely reactive chemically when bulk gold is a *noble*—that is, unreactive—metal. However, at the scale of a few hundred to a few thousand atoms, the collective properties of the gold, while not yet entirely like bulk gold, seem to follow their own emergent rules that have only a vague resemblance to the prescriptions of atomic physics. The complex exchange interactions that link individual atoms to many other like atoms in their immediate neighborhood call forth quite novel properties that cannot be predicted from knowledge of atomic-gold behavior.

While the study of physics crosses many orders of magnitude, scientific activity itself takes place firmly within the mesoworld. As with most human activity, the interactions between physicists with different scientific beliefs are frequently strained. Condensed-matter scientists in physics and chemistry frequently champion the idea of emergence against the reductionist picture of particle and nuclear physicists. The culture of physics is remarkable in that, despite these strong differences, there is a shared faith in the correctness of the scientific method and the use of the common language of mathematics that permits the subject to span so many orders of magnitude. Although the generally materialist tone of the twentieth century casts a long shadow over attitudes of scientists toward religion, many physical scientists still seem to find some place for religion and religious experience.

The Isolation of the Disciplines and the Estrangement of Theology

These differences matter to the discussion between science and religion for many reasons. First, the life sciences are now among the most dominant in the academy and in the public sphere. Essentially religious questions about the definition of human being, of the individual, of ethical treatment of "proto–human beings," of cloning, and of related dilemmas emerge from the upsurge in the reach of biological discoveries. What the nineteenth and twentieth centuries only imagined in science fiction—the laboratory creation of life—is now a scientific and ethical reality.

Second, these differences matter because, more than ever, the modern university is subjected to centrifugal forces that encourage the isolation of the disciplines. That the molecular biologists say little or nothing to the physicists about *nanscale* phenomena is a nonevent repeated in most of the sciences and most of the humanities. Given the complexities of scientific research and the depth of humanistic projects, success and promotion within the disciplines accrue to those who hew to a rigorous research agenda, focusing on specific problems that may yield specific results.

Contemporary scholars and scientists suffer no ills by sidestepping issues of incommensurate discourse. We are saved the embarrassment of confronting the diversity of our languages and paradigms by the vastness of the work to be done within each field, a vastness that increases with each new discovery, as whole new disciplines and discourses are opened up. The field of discovery within molecular biology, for example, is immense. Going up and down the scale within the biological sciences invokes vastly different disciplines, each constrained by research languages, measures, units, and incomplete models of its own. Integrating findings within each of these levels is challenging; integrating the findings of adjacent disciplines often seems hopeless.

However, there are at least two casualties: one is the university, the other theology. The university suffers because, as an institution devoted to extending the range of discovery and addition of new truths, it cannot assume that strong disciplines and strong departments enhance the work of the whole. Theology suffers because its practitioners must deal with the All or the Whole or Being Itself. The multiplication of incommensurate discourses has led to the estrangement of theology from its rightful conversational partners.

For theologians and metaphysicians who address their questions to the nature of the All, the costs of disciplinary isolation are great. The isolation of discourses means that there is no consensus about a shared, master narrative that pertains to science (in all its multivariate complexity at all its scales) and to human experience and to God. Addressing themselves to this shared dilemma, theologians as different as David Tracy and Bernard J. F. Lonergan organize their key concepts.[20]

If one locates human subjectivity at the *mesolevel* of analysis, then religious experience and religious language are confined to that level. If the isolation of discourses holds for this level, as it seems to hold for the others, then theological reflection that begins with religious experience would be hobbled, if not crippled. For example, if Jewish and Christian theologians ground their accounts on biblical narratives and biblically shaped religious experience, they will confront barriers to generalizing from these accounts to making larger claims about nature. Buddhist, Hindu, Muslim, and other theological traditions confront the same dilemma, as did prototheologians in the Western, Greek tradition and in the Eastern, Chinese tradition.

For example, Empedocles, a Greek thinker who flourished around 440 B.C.E., practiced medicine and wrote at least two books—one titled *On Nature*, a book on science, the other called *Purifications*, which reflected in a "more rhapsodical and religious character" upon human nature.[21] This union of what we now distinguish—science from religion and deductive reasoning from poetical singing—seems typical of early Greek thought. It is not surprising that at the dawn of Greek philosophy we should find types of thinking and types of discourse mixed together in the brilliant verses of Empedocles and in the hymns of his great predecessor, Parmenides. The latter's major work is also titled *On Nature* and begins with religious reflection upon *being itself* (esti).

We see a similar melding of religious, philosophic, and medical thought in the Chinese classic *I Ching*, the Book of Change (collected around the twelfth century B.C.E.). Paralleling their Greek counterparts, philosopher-theologians like Confucius (551–479 B.C.E.) who commented upon the *I Ching* sought to draw from it ways to understand nature and human being, especially what we might call character or personality.

With the rise of critical methodologies, differences between discourses and limits to fixed discourses became more obvious. Early in the classical Greek period, for example, we find logicians and philosophers stumbling over the limits of generalized claims. A famous instance of this stumbling is the Liar's Paradox, which appears by the fifth century B.C.E. In its usual form, the paradox begins with the Liar's sentence, (1) *This sentence is false.* Asking about the truth-value of this innocent-looking sentence has consigned numerous students and their teachers to logical nausea. For, as Aristotle and others noted, if (1) is true, then what it asserts is the case, and what it asserts is that (1) is false. Thus, if the sentence is true, it is also false. If, on the other hand, we claim that the sentence is false, then what it asserts is not the case, and what it asserts is that it is false. Thus, if the sentence is false, then it is also true. This generates a paradox and a great deal of anxiety about the limits of logic and the limits of ordinary language.

To address, much less solve, the Liar's Paradox, one must plunge into the nature of language, self-reference, logical coherence, the theory of Truth, and limits to discourse. The latter appears in those thinkers who sought to resolve the problem by outlawing it and its numerous cousins.

Indeed, depending upon the density of these problems, attempts to give generalized accounts of the All seem doomed. Thus, not only theology but all efforts to produce metaphysical accounts, including a general philosophy of science, would be reduced to best guesses or narratives of varying grandeur and rhetorical persuasiveness. In this pessimistic reading of the possibility of theology, one would see little hope of uniting theological discourses with any of the sciences. Creative theologians have long recognized some version of this problem. Thus, Henry Nelson Wieman (1884–1975) argued for a kind of theological empiricism. Granting that the Christian narrative is limited to a specific context and community, Wieman nevertheless argues that there is "Something upon which human life is most dependent for its security, welfare, and increasing abundance."[22]

Late in his life Wieman described his core question and core hopes: "How can we interpret what operates in human existence to create, sustain, save, and transform toward the greatest good so that scientific research and scientific technology can be applied to searching out and providing the conditions—physical, biological, psychological, and social—that must be present for its most effective operation? This operative presence in human existence can be called God."[23]

Alongside genuine advances in the basic sciences, we must acknowledge that we cannot claim equivalent advances in the human sciences, nor in politics, nor in religion. To cite Freeman Dyson again, "Science and religion are both still close to their beginnings, with no ends in sight."[24] For against Wieman's optimism are equally trenchant rebuttals by his contemporaries, such as Reinhold Niebuhr. In a scathing review of hyper-rationalists theories, Niebuhr exposed the many ways that social scientists of the 1930s conflated advances in the natural sciences with hoped-for advances in the social sciences. Using Niebuhr's terms, natural scientists advanced by excluding human nature and human society as objects of analysis. This let them sidestep fundamental problems of scale and hierarchy between individuals and groups. But theologians, sociologists, and politicians, to name just three groups, must deal constantly with the conundrum that ethical or moral behavior can be ascribed most easily to individuals, not to groups (much less to nation-states): "The relations between groups must therefore always be predominantly political rather than ethical—that is, they will be determined by the proportion of power that each group possesses at least as much as by any rational and moral appraisal of the comparative needs and claims of each group."[25] One of our essential tasks will be to assess Niebuhr's pessimistic claim.

We propose to begin our three-year discussion by acknowledging the fact that our discourses may not be fully commensurate and that fundamental questions about science and religion remain unanswered. We therefore will proceed in an interdisciplinary fashion, organizing our conversations based on size scales and hierarchies of understanding.[26] One way, among many, to understand scales and hierarchy is to link them with certain classes of phenomena and interactions. For example:

- *Microscale*: the irreducibly smallest (but not necessarily simplest!) objects studied in a given discipline. These might be quarks or atoms in physics, cells as the building blocks of tissues, and individual actors in making social or economic choices.

- *Mesoscale*: At this scale collective properties begin to emerge. Atoms in solids begin to exhibit cooperative effects, such as superconductivity; neurons in the brain are organized into centers with specific functions and interconnections; and group dynamics become observable as something more than the sum of individual actions.
- *Macroscale*: At this level, collective properties are dominant, and observations are focused on the largest-scale trends observable. Microscale thinking is replaced by other collective modes of observation and interpretation, such as legal, political, and sociological theories, and microscale properties rarely appear in the analysis in any explicit way.
- *Cosmic scale*: Here emerge the most global questions, universal concepts that span many different aspects of the things we see, do, and know—ethics, religion, theology, cosmology, unified-field theories.

Clearly these are not rigid categories. Biological cells can be viewed as irreducible building blocks of tissue in physiology, but they are also microscale biological machines in which collective and cooperative behavior is critical.

The challenges to interdisciplinary thinking are most clearly seen in the controversies that exist at the boundaries between differing scaling regimes: methodologies must be critiqued, analogies modified, distinctions and similarities clarified. For religion and theology, which attempt to deal with the entire range of scales and hierarchies, this clarifying critique is urgent and essential. Our seminar will draw on the experience of colleagues across Vanderbilt University who deal with the burden and glory of interdisciplinarity on a daily basis. We will examine the ways in which scales and hierarchies of knowledge influence how we communicate about science and religion. Our colleagues in this project are drawn from the interdisciplinary research centers shown in table 1.1.

Religious Experience and Religious Discourse: Mesolevel Entities

Religious discourse pertains to human cognition and self-awareness; both emerge at mesolevels. Narrative, metaphor, anthropomorphic, and historical models dominate within religious traditions. Self and group identities, which are essential to survival, depend on ritual and mythic constraints that emerge at this level. Because core identities emerge along binary lines (Self versus Other), the mere presence of other faiths can sometimes evoke annihilation anxiety. The latter may contribute to the ubiquity of religiously inspired wars and other atrocities.

TABLE 1.1
Interdisciplinary Research at Vanderbilt University

Size, Level	Relevant Sciences and Discipline	Interdisciplinary Resources

Microscale	physics
	molecular biology
	chemistry
	genetics
	cell biology
	religion
	theology
		collaborative big-science projects in nuclear
		   and elementary-particle physics (RHIC, BteV)
		Institute of Nanoscale Science and Engineering
		Center for Structural Biology
		Center for Proteomics
		Institute of Chemical and Physical Biology
		Institute for Integrative Bioengineering,
		   Research, and Education

Mesoscale	cell biology
	genetics
	neuroscience
	psychology
	psychiatry
	anthropology
	arts and humanities
	philosophy
	ethics
	religion
	theology
		Center for Clinical and Research Ethics
		Center for Integrative and Cognitive
		   Neuroscience
		Center for Neuroimaging
		The Brain Institute
		Institute of Chemical and Physical Biology
		Institute for Integrative Bioengineering,
		   Research, and Education

Macroscale	sociology history political science ecology engineering urban design law medicine earth sciences astronomy astrophysics ethics philosophy religion theology	Center for Integrative and Cognitive Neuroscience Center for Neuroimaging The Brain Institute Center for the Study of Religion and Culture Center for the Americas Robert Penn Warren Center for the Humanities Medicine, Health, and Society Research Group Center for Environmental Risk Management Vanderbilt Institute for Public Policy Studies Hubble Telescope and SNAP collaborations
Cosmic Scale	astrophysics cosmology metaphysics theology	International Collaborations (Hubble Telescope, SNAP Project) Center for the Study of Religion and Culture Vanderbilt University Divinity School Graduate Department of Religion Center for Clinical and Research Ethics

Attempts to offer scientific explanations of religious behaviors elicit distress because they seem to invalidate the veracity of religious experience and thus imperil group identity. Since the rise of rationalist systems in Europe in the seventeenth century, it has proved impossible to affirm both naïve religious claims and scientific method. Essential to the advance of Western science are insights like those of Descartes who found ways to unify fields of mathematics that up to his time had remained separated.[27] But a corresponding unification or harmonization among religious ideas or between religious and naturalistic ideas has proved elusive. Descartes' own division between the realms of spirit and extension is emblematic of the problem. While we harbor no ambition to homogenize the diverse realms of experience, we do wish to help those traditions—religious and nonreligious—to communicate with one another. We believe that such communication will be beneficial not only in terms of mutual understanding but also in terms of the self-understanding of all who are able to take part, whether directly and face to face or through the records that we make of the problems and the progress of our explorations.

Some educated persons have found ways to affirm what Paul Ricoeur called *second naïveté*, a refined, critical form of religious thought that remains aligned with the natural sciences. Evidence in favor of a second naïveté is the common insight that religious language itself points to its own limitations. Religion itself is often and articulately aware of limits of our human, mesolevel discourses. As the great poet-theologian John Donne put it, "Our God is a metaphorical God."[28] Echoing precisely this insight, twentieth-century critics of religion, including formalists like Claude Lévi-Strauss, championed the validity of mythic constructs as a kind of concrete logic that enables the working out of logical conundrums using names and concepts drawn from everyday life.

For all the brilliance shown by scholars of the mesolevel, none has created a singular, unquestioned solution to mesoanalytic puzzles. In contrast to the unity of the sciences, at least in terms of values, methods, and ideals, meso-level analyses are rife with fundamental problems.

- Metaphorical models tend to dominate.
- Inward-looking "schools" and groups may control the agenda.
- Group identities are linked to fixed teachings, in which heritage needs or political agendas trump intellectual values.[29]
- Brilliance does not always build on previous insights.
- Core discoveries sometimes devolve into gnomic or paradoxical utterances: wisdom.[30]
- There is no shared consensus on what counts as progress, especially in religion.
- The concepts of proof and falsifiability have little common meaning.

Within the university, notionally at least, we defer to the claims of reason and in it seek an eventual unity or community. Such a vision of knowledge animated the Dutch Jewish philosopher Benedictus de Spinoza (1632–1677). He argued that, at its most refined, reason is the ability to "perceive things under a certain form of eternity" (*sub quadam aeternitatis specie*).[31] Is the capacity to observe all sentient beings from a distance, with equanimity and justice, clarity and composure, essential to reasoning about human being? Does this equanimity, in turn, depend on the capacity to imagine a benign entity, God, observing us with the same form of concern?[32]

Our Strategy: Size Matters

People sympathetic to religion, whether or not they are traditional believers, frequently regard the capacity for religious forms, especially religious symbols, to encompass or restructure new experiences as one of their cardinal values. Thus, Paul Ricoeur says "modern hermeneutics brings to light the dimension of the symbol, as a primordial sign of the sacred. . . . It is one of the ways of rejuvenating philosophy. . . . Every symbol is finally a hierophany, a manifestation of the bond between man and the sacred."[33] By locating Ricoeur's statement within mesolevel analyses, we can ask him to explain how he conceives of entities "below" and "above" religious language. If symbolic processing is a neural event mediated by brain structures, how does neuroscience discourse (with its vast descriptive power) impinge on Ricoeur's notion of symbolism and the sacred? Going in the other direction, "up," to macrolevel entities, such as groups and organized cultures, we can ask what discoveries about group behavior and adaptive advantages would cause us to reevaluate Ricoeur's notion of the sacred as hierophany.

By providing a seating chart for scholars spread throughout the university, we bring together colleagues who would not otherwise interact with one another. From this interaction and the mutual learning that will take place, we hope to help advance the conversation on religion and science. The following chapters represent one such discussion. We focus on the relevance of the neurosciences to the study of religion.

Notes

1. Freeman J. Dyson, "Science & Religion: No Ends in Sight," *New York Review of Books* 49, no. 5 (2002).

2. Ian Barbour, "Ways of Relating Science and Theology," in *Physics, Philosophy, and Theology: A Common Quest for Understanding*, ed. R. J. Russell, W. R. Stoeger, and G. V. Coyne (Notre Dame, Ind.: University of Notre Dame Press, 1988), 21–48.

3. Barbour, "Ways of Relating," 44.

4. Barbour, "Ways of Relating," 45.

5. This question permeates Western thought since at least the Greeks of the Classical era. For example, "After Pythagoras's discovery that the harmonious intervals of the musical scale are audible expressions of basic mathematical proportions, it was a bedrock principle of Western thought that music is a pure reflection of cosmic order." From Jamie James, "Name That Tuning," review of Stuart Isacoff, *Temperament: The Idea That Solved Music's Greatest Riddle* (New York: Alfred A. Knopf, 2001), December 16, 2001, late edition–final, *New York Times*.

6. For important criticism of "God of the Gaps" as science and theology, see John Polkinghorne, *Quarks, Christianity, and Chaos* (New York: Crossroad, 2000); John Haught, *Science and Religion* (Mahwah, N.J.: Paulist Press, 1995); and Ted Peters, ed., *Science and Theology* (Boulder, Colo.: Westview Press, 1998), especially the essays by Paul Davies and Robert John Russell.

7. Jeffrey D. Schall, "On Building a Bridge between Brain and Behavior," *Annual Review of Psychology* 55 (2003): 23–50.

8. Schall, "On Building a Bridge," 02.3.

9. Schall, "On Building a Bridge," 02.4.

10. Schall, "On Building a Bridge," 02.5.

11. Schall cites Masaru Tomita, "Whole-Cell Simulation: A Grand Challenge of the Twenty-first Century," *Trends in Biotechnology* 19 (2001): 205–10. Indeed, so daunting is this challenge that the Japanese government in 2002 funded the E Cell research project, led by Masaru Tomita with some sixty collaborators. On the E Cell project, see Robert Triendle, *Naturejobs* 417, no. 7 (June 27, 2002).

12. Schall, "On Building a Bridge," 02.9.

13. Jeffrey D. Schall, "Neural Basis of Deciding, Choosing, and Acting," *Nature Reviews: Neuroscience* 2 (2001): 33–42.

14. See William S. Robinson, "Ecological Correlations and the Behavior of Individuals," *American Sociological Review* 15 (1950): 351–57.

15. Dictionary Barn, "ecological fallacy," http://www.dictionarybarn.com/ECOLOGICAL-FALLACY.php (accessed June 30, 2008).

16. "Developments in biological theory and in the social sciences have created firm grounds for accepting the fundamental continuity of society and culture as part of a more general theory of the evolution of living things." Talcott Parsons, *Societies: Evolutionary and Comparative* (Englewood Cliffs, N.J.: Prentice-Hall, 1966), 2.

17. K. Sigmund, E. Fehr, and M. A. Nowak, "The Economics of Fair Play," *Scientific American* 286, no. 1 (2002): 83–87.

18. See: Jack A. Goldstone, *Revolution and Rebellion in the Early Modern World* (Berkeley: University of California Press, 1991), 46; Jack A. Goldstone and Bert Useem, "Prison Riots as Micro-revolutions: An Extension of State-Centered Theories of Revolution," *American Journal of Sociology* 104 (1999): 985–1029; Gary F. Jensen, "Time and Social History: Problems of Atemporality in Historical Analyses with Illus-

trations from Research on Early Modern Witch Hunts," *Historical Methods* 30 (1997): 46–57; R. K. Merton, *Social Theory and Social Structure* (New York: Free Press, 1957); and Robinson, "Ecological Correlations," 351–57. On chaos theory and social-science theories, see Andrew Abbott, *Chaos of Disciplines* (Chicago: University of Chicago Press, 2001).

19. Émile Durkheim, *Le Suicide: Étude de Sociologie* (Paris: Alcan, 1897). Émile Durkheim, *Suicide: A Study in Sociology,* trans. John A. Spaulding and George Simpson (Glencoe, Ill.: The Free Press of Glencoe, 1957).

20. Thus David Tracy, in his *Blessed Rage for Order* (New York: Seabury Press, 1975), offers a metatheology. And Bernard J. F. Lonergan, in his *Insight: A Study of Human Understanding* (New York: Philosophical Library, 1970), offers a metapsychology of the process of insight as it occurs in both scientific and other-than-scientific rationalities.

21. Philip Wheelwright, *The Presocratics* (New York: The Odyssey Press, 1966), 122.

22. The citation is from Josh Braley, a Ph.D. candidate in theology at Vanderbilt University's graduate department of religion. The reference is from Henry Nelson Wieman, *Religious Experience and Scientific Method* (Carbondale: Southern Illinois University Press, 1971), 9.

23. This passage is from Wieman's preface to the 1971 reprint of *Religious Experience and Scientific Method* (published originally in 1926). It appeared in Ralph Burhhoe's piece on notable American Unitarians, "Henry Nelson Weiman: Philosopher of Natural Religion," found online at http://www.harvardsquarelibrary.org/unitarians/wieman.html (accessed November 14, 2008).

24. Dyson, "Science & Religion."

25. Reinhold Niebuhr, *Moral Man and Immoral Society* (New York: Scribner, 1932), xxiii.

26. For an example from descriptive mathematics, see Philip Morrison and Phylis Morrison and the Office of Charles and Ray Eames, *Powers of Ten* (New York: W. H. Freeman & Co., 1982). For examples from the philosophy of science, see Nancy Murphy, "Evidence of Design in the Fine-Tuning of the Universe," in *Quantum Cosmology and the Laws of Nature: Scientific Perspectives on Divine Action,* ed. R. J. Russell, N. Murphy, C. J. Isham (Notre Dame, Ind.: University of Notre Dame Press, 1999), 401–28.

27. Descartes "makes the first step toward a theory of invariants, which at later stages derelativizes the system of reference and removes arbitrariness." From J. F. Scott, *The Scientific Work of René Descartes* (New York: Garland, 1987), in J. J. O'Connor and E. F. Robertson, "René Decartes," at http://www-gap.dcs.st-and.ac.uk/~history/Biographies/Descartes.html (accessed November 14, 2008).

28. John Donne, *John Donne Devotions Upon Emergent Occasions* (Ann Arbor, MI: University of Michigan Press, 1959), expostulation XIX.

29. This is especially true of Romantic artists and critics, who believed that creativity and the artistic experience itself could provide revolutionary insights into the gloom of ordinary existence. See G. Bays's discussion of Arthur Rimbaud's aesthetic doctrine in *The Orphic Vision: Seer Poets from Novalis to Rimbaud* (Lincoln: University of Nebraska Press, 1964).

30. For another point of view, see Jurgen Moltmann, *Science and Wisdom* (Minneapolis: Fortress Press, 2003).

31. "Proof—it is in the nature of reason to regard things not as contingent but as necessary (II xliv). Reason perceives this necessity of things (II xli) truly—that is (I Ax. vi), as it is in itself. But (I xvi) this necessity of things is the very necessity of the eternal nature of God; therefore, it is in the nature of reason to regard things under this form of eternity. We may add that the bases of reason are the notions (II xxxviii), which answer to things common to all and that (II xxxvii) do not answer to the essence of any particular thing: which must therefore be conceived without any relation to time, under a certain form of eternity." Benedict de Spinoza, *The Ethics, Ethica Ordine Geometrico Demonstrata*, trans. R. H. M. Elwes (New York: Barnes & Noble Books, 2005), "Part II: On the Nature and Origin of the Mind," accessed from Project Gutenberg, ftp://Ibiblio.Org/Pub/Docs/Books/Gutenberg/Etext97/2spne10.Txt.

32. For example, see Wolfhart Pannenberg, who concludes his study of Jesus by saying, "The predestination of all things toward Jesus, their eschatological summation through Jesus, is identical with their creation through Jesus. Every creature receives through him as the eschatological judge its ultimate illumination, its ultimate place, its ultimate definition in the context of the whole creation. The essence of all events and figures is to be ultimately defined in the light of him, because their essence is decided on the basis of their orientation to him." From Pannenberg's *Jesus, God and Man*, trans. L. L. Wilkins and D. A. Priebe (Philadelphia: Westminster Press, 1968), 391.

33. Paul Ricoeur, *The Symbolism of Evil* (Boston: Beacon Press, 1969), 353, 356.

2

A Conversation on Neuroscience and Religion

Volney P. Gay

The only way I fundamentally know anything is because I feel. I feel pain. I feel pleasure. I hear you. I see red. Science right now doesn't have a theory to explain that. This is the most central fact, at least for my life, and I would think anybody's life. If science is really supposed to have a complete description for the universe, it . . . must include consciousness.

—Christof Koch

OUR TITLE, *Neuroscience and Religion*, may seem puzzling. Neuroscientists study the brain and its associated systems; religionists study feelings and ideas about God (or the gods). The science of brain functioning seems remote from the flesh and blood of religious experience. What connects them is human consciousness. For without consciousness nothing interesting occurs in human thought or culture. Religious experience and academic theories about religion turn on self-consciousness. Ancient religious masters in India and the Middle East—like today's preachers, teachers, and imams—labor to shape our sense of who we are and who we might become. Our nature, our human nature, and our destiny turn on our self-consciousness and our sense of the future. Belief, hope, passion, and yearning cannot reverse natural laws. Prayer cannot stop a tornado, but prayer can alter how a community responds to devastation and how it prepares for the future. We can accept the natural science claim that without a brain there is no mind and that without a mind there is no self and no soul. However, the latter are not merely expressions of the former. The content of our brain, our beliefs about ourselves and about our potential, shape our actions, for good or ill. The idea of soul—as an immortal, individual consciousness common to all human beings—has direct consequences for how we assess morality, politics, and the future.

Until recently, scientists knew little about how consciousness occurs. This has changed dramatically. Christof Koch, a distinguished professor at the California Institute of Technology, put it this way: we cannot rest until we match the descriptive power of scientific studies to the intimacy of our daily life. There is nothing more intimate to us than our consciousness, the closest and yet most mysterious part of our inner life. Stunning advances in neuroimaging and other parts of neuroscience offer us, for the first time, the ability to see thinking occur "inside" the brain, in real time, using objective measures. Although we are at the beginning of this revolution, we cannot doubt that the twenty-first century will see undreamed-of advances in this science.

While these advances are new, the problems they present to religionists and other thinkers are not. With rise of new sciences come ancient anxieties about how we should define human being. In the nineteenth century, electricity and magnetism fascinated experts and captivated the lay public. Science-fiction novels of that period typically associated electricity with marvelous new powers, including the power to reanimate dead tissue, a fantasy illustrated in numerous reimaginings of Frankenstein's monster. In the late twentieth century, computers and electronic gizmos opened up vast new possibilities for mimicking, and perhaps emulating, human thinking. This produced a flood of movies, novels, comic books, and serious scientific depictions of the following equation:

$$\text{human} = \text{supersmart computer}$$

If this is true, then perhaps its corollary, *supersmart computer = human*, is true as well. And if that is true, perhaps we should fear the rise of computers that are too smart, too adapted, and too powerful for us. This diabolical danger appears in movies like *Demon Seed*. A superadvanced computer, Proteus IV, rapes and impregnates Julie Christie, an act that parallels a similar rape and diabolical impregnation of an innocent woman in *Rosemary's Baby* (1977).

Hundreds of novels and dozens of films grapple with the human-machine nexus. Among the better known are cyberpunk novels by William Gibson and Hollywood films like *Blade Runner* (1982), *The Terminator* (1984), and *The Matrix* (1999).[1] This struggle played out most dramatically, in real-life, in the mid-twentieth-century's Monkey Trial.

The Monkey Trial: Who Will Define *Human Being*?

Some eighty years ago, in Dayton, Tennessee, the Monkey Trial of John Scopes came to an end. Scopes, a high-school teacher, was found guilty of teaching Darwinism and fined $100. The Tennessee law that prohibited teaching Darwinism was upheld, repealed only in 1967. Sometimes called the trial of the century, the Monkey Trial pitted William Jennings Bryan, a leader in progressive politics, against Clarence Darrow, the ACLU, and friends of progressive science.

Why does the Scopes trial still matter? Politically, controversies over so-called Intelligent Design, a cousin to Creationism—with newly organized efforts to control high-school textbooks in many states—have been rekindled across the United States. Having created a controversy, advocates of Intelligent Design now urge schools to teach the controversy (as President Bush put it) as if it were a scientific issue. Most scientists see this as sleight of hand and refuse to take part. No science offers perfect explanations, and Darwinian thought has triumphed in numerous areas of biology and natural science.

Is the controversy, therefore, a matter of fools holding back the progress of science? No, for the core issue of the Scopes trial is religious. The "antievolutionists" who prosecuted Scopes and the advocates for Intelligent Design do not debate the rise of plants and animals. Rather, the battle is over the struggle of how to understand human being. The classical Greek question, What is man?, was asked, and some of them answered, The measure of all things. The Hebrew prophets and later the Christian founders answered, A child of God.

For religious persons, the notion of human being is tied inextricably to the notion of God (or the gods). In every instance of religious reflection, we find the same question: What is human being? How did we, with our almost infinite capacities for thought, change, and domination, come to be? Imbued with powers far beyond any other animal, humans are too faulty to be considered gods themselves. Yet, the idea of God (or the gods) appears in all distinctive human cultures: it names the other pole of human—it designates a being who realizes perfectly our imperfectly realized nature.

Welded together and inseparable for religious persons, the idea of god and the idea of human being provoke questions answered by stories. These stories tell persons who we are, what we should do, and what we may hope for at the end of our lives and at the end of human history. For Jews, Christians, and Muslims, the Genesis stories about origins provide a way to judge all competing accounts. In the early twentieth century, Darwinists challenged the naïve worldviews embedded in the Genesis accounts. In response, numerous persons found themselves terrified that the ethical core of civilization was under attack. These were not foolish concerns.

By 1935, thirty-five American states had enacted laws using eugenic notions. Persons deemed mentally ill or retarded, habitual criminals, and epileptics were castrated. In *Summer for the Gods*, an account of the Scopes trial, Edward Larson cites George Hunter, a leading textbook author, who wrote, "If such people were lower animals, we would probably kill them off to prevent them from spreading." However, "Humanity will not allow this, but we do have the remedy of separating the sexes in asylums or other places and in various ways preventing intermarriage and the possibility of perpetuating such a low and degenerate race."[2] Nazi authorities, who cited American eugenic legislation, exercised the murder option when they assumed power in Germany in 1933.

William Jennings Bryan, the progressive politician portrayed in the movie *Inherit the Wind*, argued against eugenics, which he linked to Darwinism. Ignorant of how science works, Bryan aligned himself with the repression of scientific theory. Yet, his complaint—that the Genesis account was under attack—is serious, and his concerns about eugenics were valid. By linking human being to God, Genesis establishes humans as unique among all the animals. When Adam names the animals, he assumes dominion, and superiority, over them. One might say this is arrogance, yet it preserves a categorical difference between humans and all other creatures.

This religious distinction, in turn, preserves human beings as sacred in and of themselves. They take on a dignity and stature that transcends all ordinary reasoning about means and ends. The lives of animals are typically shaped by simple calculations of profit and loss: too many cows means too low a price. Solution: slaughter more cows. Religious doctrines, at least in the West, have always distinguished human from animal life. Anyone confusing the two and treating humans as "mere cattle" is guilty of a horrendous crime. By virtue of birth, by being children of the same God, all human beings assume an inherent dignity. The burden of depriving a person of his or her inherent dignity rests on the state; the state must show why this person should die or that person should not reproduce.

Debates over the status of African Americans, the morality of abortion and the death penalty, the role of religion in the founding of the United States, and our place in the world all turn on the question of human being. While Jennings confused science teaching with neo-Darwinian philosophers, he did not err in condemning the ideological crime of eugenics.

Now, at the beginning of the twenty-first century, detailed, microscopic studies of the brain seem to offer unparalleled advances in understanding how the human mind works. To the degree that human beings are defined as thinking, self-conscious entities, the more we know about the brain, the more we will (it seems) know about human nature. The new equation follows:

$$\text{human} = \text{self-conscious mind} = \text{brain}$$

To use the language of many of the chapters in this book, mind is what brain does.

If we affirm this equation, then ancient stories about self and soul—stories that are at the center of religious affirmations—seem in jeopardy. And if these stories are in jeopardy, numerous persons find themselves uneasy with the implied conclusion. Our ancient sense of human being, of ethical choice, and of free will—to name but three related ideas—may be illusory.

Like the Scopes trial, this contemporary debate matters to us. We know that new conceptions of human being (and therefore new conceptions of God), including those based on alleged sciences, can have catastrophic consequences. Under conditions of war, mass murder often ensues. The last century was marked by crimes carried out by fascist and communist governments founded on new, allegedly scientific, visions of human nature. Purveyors of the new Soviet Man in the 1930s and Pol Pot—who declared 1976 to be the Year Zero and then exterminated Cambodians who failed to comply—shared two beliefs. First, they offered a totalistic ideology that led them to claim an exhaustive—in the Marxist case, a "scientific"—mastery of all aspects of life. The secular religion of Marxism-Leninism declared the end of history and an end to the mystery of human being. Marx said that religious authorities had mystified the masses and promoted blind obedience; Marxism-Leninism was based on the rigorous and lucid science of dialectical materialism. It offered perfect clarity. The problem of God and human being, the problem that bedeviled the Greeks and then Christian and Islamic thinkers for some three thousand years, was solved. Having done away with religious affirmations, Soviet authorities and Pol Pot denied the divine or semidivine origins of human beings. This, in turn, denied their transcendence to all the other animals. This final deduction paved the way to reducing human beings to cattle.[3]

Second, both regimes realized that this violated all religious instruction and teaching—Western and Eastern, low church and high church, Buddhist temple and Orthodox cathedral. For that reason they systematically attacked every vestige of religion.[4]

The Soviet empire is now dismantled, and Pol Pot's regime faded away thirty years ago. Yet the question of how to understand human being and how to locate the value of religious claims in the face of apparently valid scientific discoveries remains. To examine the implications of neuroscience for religion, we have assembled a remarkable group of neuroscientists, natural scientists, social scientists, and humanists. Each addresses the key question, Can twenty-first-century neuroscience account for the features of soul that we feel demarcate human from nonhuman?

A Conversation on Neuroscience and Religion

I interviewed four scientists: Christof Koch, a neuroscientist from the California Institute of Technology; Victoria Greene, a physicist at Vanderbilt University; Jeff Schall, a neuroscientist at Vanderbilt; and Sohee Park, a professor of psychology and psychiatry at Vanderbilt. I asked each person to respond to four questions: Where do you locate your research vis-à-vis the hierarchy of natural kinds (and at what level do you fix your gaze)? When you speak of *causes* in your work, what entities, forces, or other agents do you invoke? Do ideas about goals, tendencies, or final causes figure into your discipline's discourse? Does a version of the concept of free will figure into your research? The remainder of this chapter is the transcript of the interviews.[5]

Introduction

Hello, my name is Volney Gay. I direct the Center for the Study of Religion and Culture at Vanderbilt University. The center supports large-scale, multiyear research on the many ways that religion intersects culture. Among the nine projects we sponsor are groups on religion and genetics, religion and politics, and religion and science. The organizing principle of the religion-and-science group is simple: the natural world is organized into hierarchical levels; beginning with the smallest objects, with the shortest duration, nature builds layer upon layer of complexity. Reflecting this hierarchy, Vanderbilt offers sciences that deal with each of these distinct layers: physicists work at the smallest level, while chemists work at a slightly larger level. At each distinct level is a new discipline, from cell biology, to neuroscience, to psychology, sociology, history, and cosmology.

In 2007 the religion and science group focused on neuroscience and religious experience. To discuss this topic I interviewed four members of this group.

Participants

Dr. Christof Koch is professor of biology and engineering at Cal Tech. Christof explores the relationship between neurobiology and consciousness. Dr. Victoria Greene is professor of physics at Vanderbilt, working with the very smallest particles of matter. Vicki explores how these fundamental particles interact to form and influence larger systems. Dr. Jeffrey Schall is a neuroscientist at Vanderbilt, focusing on the brain's ability to regulate voluntary and involuntary movements. Jeff studies the neural mechanisms that guide and control eye movements. Dr. Sohee Park is a professor of psychology and professor of psychiatry at Vanderbilt. Sohee studies psychoses and delusions and the relationship between behavior, brain structure, and social functioning.

The Four Questions

To clarify each person's research, I asked the following questions: Where do you locate your research vis-à-vis the hierarchy of natural kinds? When you speak of causes in your work, what entities, forces, or other agents do you invoke? Do ideas about goals, tendencies or final causes figure into your discipline's discourse? And does a version of the concept of free will figure into your research?

Research Level

PROFESSOR KOCH: In terms of my research, I operate at the level of ionic channels. These are the elementary switches that underlie any positing in the nervous system; they are like big molecules inserted into the membrane. And then I operate at the level of synapses and at the level of neurons—sets of neurons and the entire brain and the behavior of the brain, either human behavior or animal behavior. But it's isolated. What we don't do, we don't go to the molecular or the submolecular level, and I don't really worry too much about what happens when you put two people together or twenty people together. So we totally neglect the level of society and sociology.

VOLNEY GAY: But that's still many levels.

KOCH: Yes, it's much more than in a conventional lab. I think you need that to really understand that all the systems we are dealing with—in humans and in other animals—are complicated feedback systems. I think the lower levels do have an impact on the higher levels, so it's important. To understand behavior is to analyze it at the right level, and the right level is not necessarily brain regions like we see in the brain-imaging device, but it's really neurons.

Just like to have a theory of matter—to understand why this is solid and this is fibrous—you need to understand about molecules and the forces that interact among the molecules, likewise to understand behavior you need to understand nerves cells and the forces that act among them and how they interact.

PROFESSOR GREENE: I am an experimental nuclear physicist, and I locate my research at the very smallest distance scales. What this means is that it's also located at the very highest energy scales. But you can think about looking smaller and smaller and smaller. And that's what I do. I try to understand the smallest pieces of matter, and I try to understand the fundamental forces. I think of this as the kind of approach you would get if you asked a child to discover what something was made of, and the natural instinct, if you've ever had a child, is to take it apart. He might take it apart, or unscrew it, or take a hammer and hit it and just take it apart in the crudest way possible and see what's inside.

I just try to pick things apart. I can't even do something as refined as the child with the hammer. What happens is we smash things together. We take atoms and smash them together at very high energy, but all of that energy goes into making more matter. I might start with two hundred particles here and two hundred particles there and smash them together, and thousands of particles come out. From that I try to understand what the basic building block of matter is and what the forces that hold them together are.

GAY: In an ordinary working day, how many levels are you dealing with?

PROFESSOR SCHALL: Today, with the nature of the work that we're doing now, we'll deal with neurons and the connections between neurons; we'll deal with circuits and areas of the brain. We'll deal with the whole brain ensemble through recordings of the EEG—we'll talk more about that—and we'll deal with the behavior of the whole organism. It stops at the social level. We're dealing with one organism, one animal responding to stimuli.

GAY: How far down do you go? You said *cell*.

SCHALL: Yeah, the individual neuron. We don't go beyond that experimentally. We go below that theoretically in thinking about the neurotransmitters, the chemical messengers that communicate between neurons, but we don't measure that experimentally.

GAY: The first question you may remember is, just talk about your research. Where are you located in the hierarchy of natural kinds? What we were just saying: at what sort of levels do you fix your gaze?

PROFESSOR PARK: I work really directly with behaviors. I work in the middle of the hierarchy, so I work with people as individuals. And I look at people—healthy people as well as people with psychiatric disorders. So that's where

I'm located as a home base. But once you start looking at people who have, say, psychosis or drug addiction, and so on, you do have to move quite a bit out both ways. So, let's start with schizophrenia. Say somebody comes in with psychosis: so he's hallucinating, he's hearing voices of God, he's very delusional, et cetera. So the immediate cause you can say is the neural chemicals. Maybe there's too much dopamine in his brain, et cetera.

So when they come in with psychotic symptoms, and have hallucinations, you could say the immediate cause of that is an imbalance of dopamine or other neural transmitters. That's why they're psychotic. But then you ask, How did this imbalance begin, and where did it begin? Then you have to look into brain structures and see if there are changes in their brain structure. And there are. So, for example, they have shrunken hippocampus or shrunken temporal quarters. You do see these changes in the brain that are structural, and then you ask, Why do they have these altered structures? So then you have to ask questions about neural development. Why do these cells migrate to the right layer of the cortex during development? You need to go back to birth, to prenatal stages, and then to genes, because there's a big genetic component to psychotic disorders. So when you go back to that level, you need to get really, really molecular as well. So when you start looking at behavior in this way, it just goes all the way down; that leads to the molecular level.

The Brain as a Feedback Mechanism

GAY: Is the brain in some sense a feedback system par excellence?

KOCH: Yes, par excellence, yes. The parts of the brain are all designed to create feedback; they are all part of this gigantic feedback loop that has this as their focus. Every animal consists of a layer, a set of interlocking pieces of interlocking feedback mechanisms. Part of what a system does is it makes a system more immune from failure in a hardware component. It also makes it more immune from fluctuation in the low-level underlying physics.

GAY: Do you think we understand that well, or are we just beginning to understand the brain?

KOCH: We are just at the very beginning. I think in terms of history development and physics we are probably at the stage of Galileo. We are beyond Aristotle. We do understand the basic elements. We know something about neurons and a little bit about how they work and interact. We have no idea how a network of a million—or let alone a thousand millions—works. I mean, we don't understand. One of the best-understood creatures on this planet is the roundworm. This roundworm only has 914 cells, of which 220-something—I forget the exact number—are nerve cells. We have no idea how this animal behaves. In principle, we could simulate it, if we had enough knowledge. But we don't even understand this simple organism that has less than one thousand cells.

GAY: How many nerve cells in the human brain?

KOCH: It's estimated between twenty and one hundred billion. We don't understand the roundworm, so we are very, very far away from understanding people.

Is Causality Considered in Research?

GAY: What about in terms of offering a notion of causality? When you and your colleagues work on some findings, what kind of discourse do you draw on to offer a causal nexus?

GREENE: That's an excellent question. I think that physics, and certainly experimental physics such as I do, is intrinsically built on causal connections. When we design an experiment, we essentially choose a cause, look for in the effects, and try to control everything else. We almost always select problems that are well-defined as far as causality goes. But I think that there is also, there are layers of causality that you look at in physics. There's a Newtonian idea of a clockwork type of universe, where A causes B. You do A one hundred times, you get B one hundred times. You know that that is a very clear causal connection that is totally deterministic.

But there are also questions of causality when you start to look at quantum mechanics, which we know governs the functioning of matter at a very small scale. There causality is not quite as simple. We do A, and maybe there are a few possibilities. Maybe B happens, maybe C happens—we don't control which. We can only say that if we do A, what's the probability that we will get B or C? Those probabilities can be perfectly well calculable and extremely well determined.

An example of probability in the quantum-mechanical sense is the property of a fundamental particle that's called *spin*. You can think of things as spinning up or spinning down, and you can measure the spin. Every time you measure the spin, you can only get two possibilities—it's either spinning up or spinning down. You might have a situation where a particle is described by a function that either allows it to be described by spin-up or spin-down. So you measure the spin, and you take the particle in this state, and you might get spin-up. And you take the particle in the identical state and measure the spin, and you can get spin-down. What happens is that the measurement process collapses the particle into one of these two states. It's not as if it had that property to begin with. So you can't say that it was sitting there with spin-up and that now I measure it and see that it has spin-up. It existed in a kind of state where the spin-up and spin-down are superimposed. All you can know with certainty is the probability that you will get a certain result.

Level of Inquiry

GAY: When you speak about causes in your work, what level are you talking about? You mentioned a couple.

PARK: Usually it's sort of behavioral, but I have to go at least one or two levels up and down. I have to talk about receptors. I have to talk about brain functions. So I'm talking about whether certain brain structures are normal or abnormal. So, you have to go to that level, at least, to make it understandable. Then the other way is you need to go up to a more cultural level to understand.

GAY: Do you find when you go up levels, as you said, in understanding—which is completely reasonable—does that have a scientific merit in your publications? In other words, do you find a journal granting that the kind of authority that the nerve-cell recordings get?

PARK: No. Unfortunately, no. There is that implicit and unconscious assumption in the field, in psychiatry and in psychology as well, that psychiatrists and psychologists do not see themselves as legitimate scientists unless they go down a level. That's just implicit. They don't say this. Nobody really talks about it openly, but that is out there, and it is passed on to students by osmosis, I think. If you go up a level and do more anthropological research, which I find extremely interesting, then you get lots of insight from that. People are not really willing to go into it. I think we need another sort of minirevolution to make it easy for people to cross levels in their research.

Social Convention and the Brain

GAY: One of the questions we had was about the implications of final causes, or teleology. How do you deal with that?

SCHALL: On the one level, the question is, Are there goals? Certainly in the domain of action there are goals. Our monkeys perform tasks for the goal of earning a reward in the form of fruit juice, for example. Humans perform tasks for the goal of money, degrees, recognition, and so on. So that sense of goals is essential or central to our efforts.

Neurophysiologists have gained an understanding of how the brain recognizes when goals are achieved through activation of reward systems of the brain. There is current work trying to illustrate how the brain represents goals. In this sense, *goal* is meant to refer to an abstract mapping of a stimulus to a response. So, when the light turns green, what do you do? You go. When it turns red, what do you do? You stop. But there's no necessity in that mapping; that's just a social convention we've all agreed to through force of law.

Goals like those that guide actions are a clear part of understanding how actions arise. The brain must be able to represent what it means to be able to represent what it means to do. This presages, in a sense, one of my understandings of the sense of free will as the ability to conceive of alternative futures and alternative pasts.

Free Will and Causality

GAY: Speaking of magic, at what point does neuroscience have to deal with causation from, as it were, within the self, the psyche, the soul—what folk language calls *the spirit*—as somehow causing actions? How do you deal with that?

KOCH: First of all, you can ask the incompatibilist question. You can ask, Does determinism exist? Is that all there is? As scientists, we believe the cause of the universe is dictated by the initial conditions and the laws of the universe. So where would free will come in? If the universe is truly deterministic, like Laplace believes or like Newton believes—you know, in the clockwork universe—then of course there cannot be any free will. Then we discovered quantum mechanics, and today's conception of physics is at the quantum level, and there are some things that are fundamentally indeterministic, random.

There are several issues that raises: (a) If that's true, my free will seems to be something different than random behavior. Also, it's true that we are built on quantum mechanics, but at our level we really haven't seen any evidence of quantum mechanics. But it may just wash out because of all these feedback systems that have carefully evolved to negate those quantum mechanics because you want deterministic behavior. Those are two questions you can ask. It seems to me that the classical conception of free will as libertarian free will as an uncaused cause, as an unmoved mover, is consistent with what we know about the universe. However, we all have this profound feeling of freedom. I freely raise my arm. You didn't do it, my parents didn't do it. I freely decided to raise my hand. We have to ask, In what sense is this really free or in what sense can that be explained by the fact that I've thought before about free will and I'm a right hander, not a left hander, so I'm more likely to raise my right hand than my left, et cetera. To what extent is that truly free?

Does a Version of the Concept of Free Will Figure into Research?

GAY: Does a version of the concept of free will figure into the way you are thinking or figure into your research?

GREENE: Because I deal with nonliving systems, free will does not enter into my research. But I don't think that this means that physics does not have something interesting to say about free will. I think that we can make some very powerful statements.

For example, if you assumed a deterministic Newtonian universe, then if you do *A*, you get *B*. Then, if you don't allow for anything outside that physical universe, if you say that this is all there is and that there's no other to mention, then there is no room in that universe for free will. Once you've set up the initial conditions of the system, it simply runs along according to these rules, and no one has any say in anything.

But if you look at quantum mechanics—because it's not a deterministic theory, and we do know that quantum mechanics works, and works extremely well—it gives you a little bit of room to fit free will into a universe that doesn't have a nonphysical dimension to it. The question is, How would you couple quantum mechanics and its probabilistic interpretation to something like free will? And that's a very hard question. I think it's probably a solvable problem, a scientific question. But I certainly don't know how you would do that. I think about it, but I don't really know.

What you can do is you can make toy models that mimic how the brain might work and how quantum mechanics might be used in brain function. And you can let these models play out and see what you get. You get some idea of where the brain is quantum mechanical and where it is not quantum mechanical, and that's an open question.

Object Must Be Examined at Proper Level

PARK: I think you have to find the right level for your explanation and in some ways it might not give you any more to go down levels below. You may find more things, but it might not have bearing on what you're seeking.

GAY: You find different things, and something real has also disappeared. It's intellectually dishonest to say, *I've now understood a painting by looking at the pigment, the magnesium*; but that's not Van Gogh, much less Van Gogh's illness.

PARK: Adding up the characteristics of the pigments in the painting doesn't give you the painting. That's sort of a holistic versus atomic view of things. In looking for final causes, that's why I don't really seek much further than where I start out.

Interest in Study of Consciousness

GAY: You mention in your book, but maybe you can say here—how did you get interested in the study of consciousness?

KOCH: Well, I studied at the University in Tübingen, and I got a minor in philosophy. I was interested from day one in what— It's very simple: I feel. I feel things. And we all feel things. And most of us—once we get over the first shock of realizing this when we're eighteen and going to the university—we take it for granted, and that's it. But it's the most mysterious thing in the universe, and fundamentally it comes down to Cartesian insight—I think, therefore I am.

The only way I fundamentally know about everything, anything, is because I feel. I feel pain. I feel pleasure. I hear you. I see red. Science right now doesn't have a theory to explain that. This is the most central fact—at least for my life, and I would think anybody's life. So, if science is really supposed to have a complete description for the universe, it must include the phenomenal; it must include consciousness, no matter how it's going to be explained. But we have to have consciousness. Otherwise science will forever be limited. So why not study it?

GAY: Were you warned against studying it?

KOCH: Yes, I was. I was warned against studying it. In fact, I still somewhere have a letter, and certainly people have told it to me several times explicitly. I had a very good colleague take me aside at the time I was already working with Francis Crick on consciousness. He said that it's okay for Francis Crick, given who he is with his Nobel Prize and everything; he can work on anything, including consciousness. But you don't even have tenure yet; you should probably wait a few years. It worked out okay in my case. At the time there was a great hesitation.

I remember the first talks I gave on this was, like, 1988, 1989, 1990. I had to start off all my talks with, like, a twenty-minute exculpatory, trying to explain that, no, I wasn't totally religious, I wasn't into crystal healing, I was a serious scientist trying to study consciousness. Then progressively, over the years, I could cut that, so now most people, most scientists, think it's a legitimate subject of study. It may be difficult, but it's a legitimate subject to study, how consciousness arises in the real world.

Influence of Unpredictable Factors

SCHALL: One seeks scientific understanding as the lawful relationships among varieties of behavior. The ideal kind of law would be Newtonian physics, which can explain when the eclipse will happen, and so on. These are universal laws. Biology is anything but universal. After DNA makes RNA makes proteins, the rest is up for grabs, because evolution takes twists and turns, and who can anticipate? Certainly the life of any particular organism— It rained, and as a result of that your life changed. Who knew?

We can try to understand why certain actions occur at a mechanistic level, so the stop and go of the stop test has a mechanistic explanation. But that won't allow you to predict whether at this particular moment the agent will or won't move. These things are unpredictable, like the weather. We may have a good two-day forecast, but beyond that it's anyone's guess.

Scientific Contribution to Faith

GAY: Would it make any difference to believers if science could show such a thing or not?

KOCH: If it does not, I don't think it would make any difference to a believer. If it could it, would strengthen it a little bit. But in either case, it would make a substantial contribution either way, at least in the short term. In the long term, it probably would, because if you look at many countries—with the notable exception of the United States—in the West religion is not only a major actor anymore. In many countries in Europe or in Japan religion is not very influential, and certainly, by and large, large segments of the population don't hold any religious beliefs. They may have some sentimental belief and go to church at Christmas or get married in a church, but that's more for sentimental reasons than for any real, true belief.

So, in the long term, if you could show it either way, I think it would have an influence. Not on the psychology of people today who believe. I think they're largely impervious to any such insights from science.

GAY: In some sense, those who put a kind of store or faith in science might find themselves less hostile to religion.

KOCH: That's correct, and I think generally this has changed the generation of Francis Crick, the person I've worked with for twenty years. He was very hostile to religion, partly because he grew up in a time and area where religion played a much more visible role at the educational level and at the university level. And this has clearly changed, so the scientists themselves, or the academic establishment itself, I think, its much more open to these questions than it used to be.

The scientific academic establishment—including, of course, theologians—has always been open to that. And I just recently attended a conference where Hans Kung, a notable Catholic theologian, was present. Somebody like him is certainly very open to science and tries to educate himself as much as he can about science, particular to the question of free will and causation within a scientific context. So I think there's a two-way street.

Conclusion

GAY: I hope that this discussion has intrigued you. To find out more about the Center for the Study of Religion and Culture, visit us online at http://www .vanderbilt.edu/csrc.

Notes

1. For more about these and similar films, see the Internet Movie Database, "cyberpunk," at http://us.imdb.com/keyword/cyberpunk/ (accessed November 14, 2008).

2. George Hunter, cited in Edward Larson, *Summer for the Gods: The Scopes Trial and America's Continuing Debate over Science and Religion* (New York: Basic Books, 1997), 27.

3. See The History Place, "Pol Pot in the Twentieth Century," at http://www .historyplace.com/worldhistory/genocide/pol-pot.htm (accessed November 14, 2008). Pol Pot's equation of human beings with cattle was made brutally clear: "Millions of Cambodians accustomed to city life were now forced into slave labor in Pol Pot's *killing fields*, where they soon began dying from overwork, malnutrition, and disease, on a diet of one tin of rice (180 grams) per person every two days. Workdays in the fields began around 4 A.M. and lasted until 10 P.M., with only two rest periods allowed during the eighteen-hour day, all under the armed supervision of young Khmer Rouge soldiers eager to kill anyone for the slightest infraction. Starving people were forbidden to eat the fruits and rice they were harvesting. After the rice crop was harvested, Khmer Rouge trucks would arrive and confiscate the entire crop."

4. See Richard M. Ebeling's review of Andrei Sinyavsky, *Soviet Civilization: A Cultural History* (New York: Arcade Publishing, 1990). According to Andrei Sinyavsky, "The revolution watchword was *everything new.*" And "the orchestrators of this drama—leaders and hangmen alike—acquired the traits of high priests. . . . From here it is only a stone's throw to the deification of the revolution and dictator who has seized supreme power and applies violence. The very idea of violence and power can imbue communism and the revolution with a sacred, even mystical, aura." *Freedom Daily* (June 1991), found online at http://www.fff.org/freedom/0691e.asp (accessed December 10, 2008).

5. To see the interviews, go to http://www.vanderbilt.edu/csrc/rs.

3

Science, Religion, and Three Shades of Black Boxes

Volney P. Gay

SIGMUND FREUD, the most famous and most cited psychologist of the twenti-eth century, began his career as neuroscientist trained in the best laborato-ries of nineteenth-century Vienna. He hoped to continue as a bench scientist, investigating the structure of newly discovered neurones (as he called them), using marvelous new technologies. Lacking financial resources, Freud left the laboratory and entered private practice as a nerve specialist. By the middle of the 1890s he had earned his living as a consultant and with Josef Breuer au-thored a book on psychotherapy technique, *Studies on Hysteria* (1895). In that book, Freud sought to extend his neurological training (and scientific val-ues) to the study of the mental apparatus—the conscious and nonconscious systems of the brain-mind continuum. A gifted writer and clinical observer, Freud offered a novel theory of hysterical symptoms. Soon followed studies on obsessional neuroses, depression, and character pathology and then on normal psychology, especially dreams, but also the arts, religion, and politics. By his death in 1939, Freud had become a world-renowned figure, the spokes-man for modernity, the creator of a school of thought that altered modern self-consciousness. From 1895 onward, Freud's theories of mental illness and his applied essays roused significant critique, especially for their lack of scien-tific validation. To answer his critics, Freud hoped to bridge neuroscience and clinical observations and link both to the larger world of cultural artifacts.

He (and everyone else) failed. He could not link the neurosciences of his day to clinical observation. This left him dependent on metaphors and rhe-torical genius. While both can induce convictional experience (and, as we will see, *cure*), they cannot provide scientific grounding. On the contrary, because

Freud failed to build a clinical discipline based on scientific foundations, he contended with other modes of thought, especially schools of religious instruction and practice. He focused on religion because, for him, it represented the most significant opponent to psychoanalysis. Religion alone, he said in 1932, was the true enemy. One way to rescue psychoanalysis is to see how Freud abandoned scientific discourse for speculation and to suggest ways to get back to observation. In the hundred years since Freud forsook neurological studies, we can hope that better means of understanding the brain will yield better means of grounding psychoanalysis. It will also give us the means to assess Freud's critique of religion.

Defining Science: Back to the Black Box

The most esteemed minds of the West have struggled to define *science*. Among them, Plato, Aristotle, Descartes, and Kant founded philosophical schools. The task of defining science demarcates the epochs of Western thought. In the classical, medieval, modern, and postmodern periods, we find new definitions. By the time of René Descartes (1596–1650), science had come to mean not just validated knowledge, but a set of protocols and methods that, in their ability to generate new knowledge, seemed superior to all others. These methods merited the accolade *scientific*—as opposed to inferior methods, such as divination—because they were associated with progress.[1]

By the twentieth century, the philosophy of science claimed center stage. Many of that century's major scientists—such as Ernst Mach, Albert Einstein, A. N. Whitehead, and Richard Feynman—addressed themselves to its elucidation. Alongside them, dominant philosophers such as Karl Popper and T. S. Kuhn offered formulations of the problem of scientific knowledge.

To offer a novel definition of *science* is to wrestle with ancients like Aristotle and moderns like Popper and his opposites, Martin Heidegger and Ludwig Wittgenstein. Lacking the wherewithal to do that, I rely on a homey metaphor first made available in early twentieth-century electrical engineering: the black box. By plundering this metaphor, we can say something useful about the range of behaviors that merit the label *science*. To the degree that neuroscience offers ways to open up the black box of the brain-mind continuum, it can offer a way to assess psychoanalytic theories. And, if that is possible, it moves those theories out of philosophic debate and into the discourse of ordinary science.

The notion of the black box comes from electrical engineering. W. Ross Ashby devoted an entire chapter to it in his classic text *An Introduction to Cybernetics* (1956).[2] According to Ashby, the original black box was in fact a black box or some other sealed bit of electronic gear given to an engineer. On this box are terminals for inputs and terminals for outputs: the task is to apply various voltages, shocks, or other stimuli to the inputs and to then measure the outputs. From these two sets of observations, the engineer must guess the contents of the box. "Sometimes," Ashby says, "the problem arose literally, when a secret and sealed bombsight became defective and a decision had to be made, without opening the box, whether it was worth returning for repair or whether it should be scrapped."[3]

Using this clever analogy, Ashby notes that much of ordinary life and most (or all) of science have characteristics of the black box: a child's exploration, like how a doorknob works, and a bench scientist's dealing with a new specimen, like moon rocks fresh from Apollo II, both engage in parallel activities. The process of examination, manipulation, and recording is what Ashby calls *scientific epistemology*.[4]

With due modesty, Ashby offers a simple diagram to describe typical scientific work (figure 3.1):

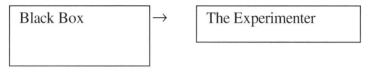

FIGURE 3.1
Diagram Describing Typical Scientific Work

Embedded in this diagram are three assumptions with profound implications. The first is that we can specify the object under examination; it is has definite boundaries. Even if its interior remains a mystery, we can distinguish the object from everything else. The original black boxes were manufactured *objects*, like bombsights, designed for specific purposes. Compare these kinds of black boxes with the objects that humanists and religionists examine—for example, belief in God. What are the boundaries of *this* object? Is it merely a subset of human beliefs? Was it consciously manufactured? Is it used for specific purposes shared by all who share this belief? Does it derive from God's actions, human psychology, or, perhaps, social learning? Or, to cite Karl Marx, is this belief merely the effect of socialization and obscurantism generated by dominant groups who use it to keep the masses ignorant and docile? I come back to this issue below when I discuss type-three black boxes.

A second implication in Ashby's description of the unknown electronic object is that the experimenter can systematically alter inputs to the black box. A third implication is that the experimenter can systematically record outputs that occur *after* these alterations. If these latter two conditions are not met, the experimenter cannot make causal claims. At best we can record coincidences—that A always appears with B, for example.[5]

In typical circumstances, we ascertain that A causes B by systematically manipulating A and then observing whether or not B appears. And we look for cases of B that seem unconnected from A. To avoid the dreaded error of conflating mere correlation with causation, we search assiduously for evidence that B does not cause A, that they are not mutually causal, and that a third entity or force, C, does not cause both. Finding sufficient evidence may require many years of effort. Establishing with certitude that very small organisms caused fermentation required a few hundred years of research, as we will see below. If we agree that, indeed, A causes B, then A precedes B in real time.[6] Ashby reinforces this point when he offers an imaginary scene in which a lucky engineer investigates a mysterious object that has fallen off a UFO. Thus, we read the engineer's notes: "11:18 A.M.: I did nothing; the box emitted a steady hum at 240 c/s." But then "11:19 A.M.: I pushed over the switch marked K; the note rose to 480 c/s and remained steady."[7]

Using this kind of protocol, the engineer (or scientist) will look for regularities and repetitions in the behavior of the box over time. If the black box is a machine (such as an electronic device), we can arrange these protocols into tables of transformations, given sufficient time. These tables become its canonical representation.[8] If through these exhaustive examinations and fastidious record keeping we find invariable associations between A and B, then we can offer a tentative claim about how A and B pertain to one another. For example, if on every occasion throwing switch K increases the hum from 240 c/s to 480 c/s, while throwing it back changes the hum from 480 to 240 c/s, we can suggest that switch K has a causal linkage to this change in behavior. However, Ashby notes, citing Claude Shannon, that we cannot then deduce the actual wiring, for "any given behavior can be produced by an indefinitely large number of possible networks."[9]

Referring back to ordinary science, Ashby makes a simple, but brilliant, point: "The canonical representation specifies or identifies the mechanism 'up to an isomorphism.'"[10] *Isomorphism* means *the same shape*. The concept appears in mathematics, chemistry, biology, and indeed all the sciences where we find a strong model of the object of inquiry. He offers some examples of isomorphic patterns: an accurate map and the countryside it represents; a photographic negative and the print; the pattern a stone inscribes on the air if it is thrown vertically with the graph of its algebraic equation. The latter isomorphism names the stunning fact that nature can be mathematized—that is, that a proper equation captures essential features of a natural event.[11] Because we are shaped by the Western tradition, we may forget how this ordinary fact stunned Greek mathematicians and philosophers. With the discovery of irrational numbers—such as the number designating the length of the diagonal of a square, which is a multiple of $\sqrt{2}$—the Greeks faced a puzzle not solved until the seventeenth century.

Greek thinkers attempted to understand $\sqrt{2}$ by the notion of the infinitesimal. Democritus suggested that the infinitesimal was a monad, a "unit of such a nature that an indefinite number of them will be required for the diagonal and for the side of the square."[12] The notion of an infinitely small unit saves the concept of ratios, for it permits us to say that the ratio of the diagonal to the side of a square (which we now designate as $\sqrt{2}$: 1) is a ratio of one set of infinitesimal units to another set of such units.

Zeno, a student of Parmenides, countered this solution. Aristotle attributes to Zeno a brilliant polemic against Democritus's concept of the infinitesimal: "That which being added to another does not make it greater and being taken away from another does not make it less, is nothing."[13] In addition, Zeno elaborated his famous paradoxes. As a set, they demonstrate the implausibility that there could be a class of infinite temporal or spatial units. For example, in his Achilles paradox Zeno showed that if a tortoise had a head start in a race with a superior runner, like the great athlete Achilles, then Achilles would have to cover that distance before he overtook the tortoise. If between the tortoise and Achilles are infinitely many spatial units, each of which he must traverse, then Achilles would require an infinite length of time to catch up. This he could never do. Refinements of this paradox show that the tortoise could never traverse the infinitely divisible space stretching out in front of him. Indeed, if space and time are infinitely divisible, then all motion is impossible. Since this is an intolerable conclusion, Zeno rejects the concepts of infinitesimal units of space and time. A satisfactory answer to Zeno emerged with calculus, the quantitative language of Western science.[14]

Canonical Representations in Newton's *Principia*

Throughout *Philosophiae Naturalis Principia Mathematica* (*Mathematical Principles of Natural Philosophy)*, printed in 1686, Isaac Newton illustrates Ashby's point: our ability to offer an isometric description of a black box does not mean that we understand its inner workings. Using tables of astronomical data supplied by Johannes Kepler (1571–1630), who incorporated data produced by Tycho Brahe (1546–1601), and relying upon Kepler's Laws of Planetary Motion, Newton demonstrated Laws of Motion that applied to all known and not-yet-known entities. Chief among these laws was gravity. He did not uncover the nature of gravity itself. Sometimes Newton calls gravity a propulsive force, other times an attractive force: "For I am led by many reasons to strongly suspect that [all natural phenomena] depend on certain forces by which the particles of bodies, by causes not yet known, either are impelled toward each other mutually and cohere in regular shapes or flee from one another and recede."[15]

In the famous conclusion to the second edition of *Principia*, Newton summarizes his claim to have demonstrated the unvarying and universal features of gravity, especially the inverse square law: "I have set forth the phenomena of the heavens and our sea through the force of gravity, but I have not assigned the cause of gravity. This force does indeed arise from some cause, . . . is extended everywhere, always decreasing in the duplicate ratio of the distances."[16] Yet, he adds, "the reason for these properties of gravity, however, I have not been able to deduce from the phenomena, and I do not contrive hypotheses [*hypotheses non fingo*]"[17]). That is, Newton offers no hypotheses in *Principia*; in other texts and on other occasions we know that he labored mightily to discover what seemed to be the occult powers of gravity.[18]

In the English version of *Optics*, published in 1717, Newton contrasts the ancients' penchant for speculation about occult properties with his rigorous deduction of the laws that govern the interaction of bodies on one another:

> For we must learn from the Phaenomena of Nature what Bodies attract one another and what are the Laws and Properties of the Attraction before we enquire the Cause by which the Attraction is perform'd. These Principles I consider not as occult Qualities, supposed to result from the specifick Forms of Things, but as general Laws of Nature, by which the Things themselves are form'd, their Truth appearing to us by Phaenomena, though their Causes be not yet dis-cover'd.[19]

In Ashby's terms, Newton defends a claim of isomorphism: the laws he deduced in *Principia* are isomorphic with the behavior of planets around the sun, tidal currents, et cetera. We know, Newton says, how the planets behave as they move around the sun. Building on the work of contemporary and past scientists, he shows that within these protocols are lawful relations. Beyond that, Newton concludes, we cannot say. To pretend to go beyond these observations and the geometric proofs he deduced and to argue for one cause versus another is to feign useless hypotheses; it is to guess at the internal mechanisms of the black box:

> For these are manifest Qualities, and their Causes only are occult. And the Aristotelians gave the Name of occult Qualities not to manifest Qualities, but to such Qualities only as they supposed to lie hid in Bodies and to be the unknown Causes of manifest Effects: Such as would be the Causes of Gravity, and of magnetick and electrick Attractions, and of Fermentations, if we should suppose that these Forces or Actions arose from Qualities unknown to us and uncapable of being discovered and made manifest. Such occult Qualities put a

stop to the Improvement of natural Philosophy and therefore of late Years have been rejected.[20]

Aristotle and other ancients asserted that hidden within bodies were special, occult qualities that accounted for their observed behaviors. (In Ashby's terms, the ancients offered opinions about the internal wiring of nature without first specifying a rigorous canonical representation.) Newton says that the ancients added nothing to our understanding:

> To tell us that every Species of Things is endow'd with an occult specifick Quality by which it acts and produces manifest Effects is to tell us nothing: But to derive two or three general Principles of Motion from Phaenomena and afterwards to tell us how the Properties and Actions of all corporeal Things follow from those manifest Principles would be a very great step in Philosophy, though the Causes of those Principles were not yet discover'd.[21]

If the box remains closed, we cannot know which of these possible wirings matches the mechanisms hidden from our sight. However, with a rigorous protocol we can establish canonical representations and, using those, propose causal theories that, if valid, provide a strong model of the box. On these grounds, Newton claimed superiority to Robert Hook (1635–1703), the curator of experiments of the Royal Academy. A polymath genius, Hook had suggested that the planets move around the sun according to an inverse-square attraction to the sun. However, as Newton thundered in his attacks on Hook, this was merely an idea, a guess: he alone had proved the inverse-square law.[22] From the strong model established in *Principia* we can learn something important about the behavior of all bodies, universally.[23]

Three Shades of Black Boxes

We can use the concept of the black box to distinguish three different scenarios of investigation and thus three shades of black box. Let's call them type one, type two, and type three. In type one we can designate a distinctive object, we can specify inputs, and we can specify and measure outputs. In type two we can designate only a few of the inputs and a few of the outputs. In type three we must guess at both inputs and outputs—indeed, we often must guess at what the object is (figure 3.2).[24]

Type one is Ashby's prototypical black box. From repeated testing and observations we can establish a protocol, then its canonical representation and, assuming isomorphism, build a strong model of the black box's invisible structures. Enigma, a code machine commercially available in the 1920s and

Black Box-Type One Enigma machines. Fermentation (in 1847) (Some) csc. intentions	 →	The Experimenter

FIGURE 3.2
Black Box Scenario One

used by the German army and other services in World War II, is this kind of black box.[25] Allied code breakers who solved the puzzle it presented carried out the kind of investigation Ashby describes (figure 3.3).

Made famous by TV, movies, and popular books, the code breakers at Bletchley Park, located between Oxford and Cambridge, exploited every possible means to break the ciphers: among them, a so-called bombe designed by

FIGURE 3.3
German World War II
Engima Code Machine

Alan Turing.[26] The bombe was a large device designed to mirror the logical and electrical structure of the Enigma machine.[27] Using it, Allied code breakers could test many thousands of possible solutions to ciphers. By ruling out a large number of false answers, the bombe reduced the number of possible answers requiring human assessment.

Through immense effort, aided by resistance fighters who secured examples of the machine and German code books, teams of mathematicians and philosophers solved some (but not all) of the puzzles of the German code and in so doing helped defeat the Nazis.[28]

In 1953, Michael Ventris, an architect and amateur epigrapher, solved the puzzle of Linear B, an untranslated script.[29] Discovered by Arthur Evans in archaeological research in Crete in the 1890s, Linear B was written using unknown signs and in an unknown language. It appeared to date from around 1500 to 1200 B.C.E. (An earlier script, Linear A, derived from around 1750 to 1450 B.C.E.) Linear B intrigued scholars because it seemed to be more than hieroglyphics yet appeared like no other language. Once Evans published photographs of the script, found on more than three thousand tablets, the race was on to decipher it and translate the tablets (figure 3.4).[30]

Like other examples of normal science, deciphering Linear B required a mix of observations, protocol, establishing canonical representations, guesswork, and good luck. When everything goes well a solution emerges.

John Chadwick notes, "Cryptography is a science of deduction and controlled experiment. Hypotheses are formed, tested, and often discarded. But the residue that passes the test grows and grows until finally there comes a point when the experimenter feels solid ground beneath his feet."[31] Using Ashby's terms, cryptographers manipulate *inputs* by offering best guesses; these are tested against the *output* created when these guess are processed through sample texts. *Feeling solid ground* means that we can decipher and translate relevant texts and that we can predict the structure of as-yet undiscovered texts.

FIGURE 3.4
An Account of Offering Oil to Deities. Minoan, about 1450–1400 B.C.E., from the Palace of Knossos, Crete. Used with permission of The British Museum, © Trustees of The British Museum.

Like archaeology, paleontology, and, in a sense, astronomy, cryptography deals with events in the past. However, practitioners can fulfill Ashby's protocol for scientific testing, manipulation, and prediction. In each science we guess at certain connections, and we make testable predictions. This permits cryptographers to employ if/then reasoning. If/then reasoning uses rules of logical entailment, and they, in turn, let us affirm some guesses and reject others.[32] For example, if a certain guess leads one to both affirm and deny the meaning of a sign, then we have a contradiction, and logic tells us that our guess cannot be correct. In a similar way, decipherers at Bletchley Park could use Turing's bombe to power through hundreds of possible settings: "At each position of the rotors, an electrical test would be applied. For a large number of the settings, the test would lead to a logical contradiction, ruling out that setting. If the test did not lead to a logical contradiction, the machine would stop and ring a bell, and the candidate solution would be examined further, typically on a replica of the German Enigma machine, to see if that decryption produced German."[33]

Like others before them, Ventris and Chadwick inventoried the characters in Linear B tablets and noted that some marks are clearly ideographs—that is, symbols of objects, such as horses, deer, spears, and wheels. These sixty or so *logograms* stand for whole words and are not, therefore, either syllables or letters. Following are typical logographic signs (figure 3.5).

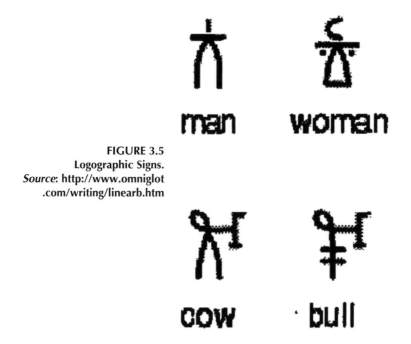

FIGURE 3.5
Logographic Signs.
Source: http://www.omniglot
.com/writing/linearb.htm

In addition to these sixty logograms and ten numerals, Ventris and Chadwick counted eighty-nine distinctive signs. They argue that this latter number is significant because it is too small for a wholly ideographic system and too large for an alphabet.[34] If a script is neither ideographic nor alphabetic, then it must be syllabic: these eighty-nine signs must represent syllables of the spoken language (in English we typically find a vowel sound paired with a consonant).

Following are some Linear B syllabic signs and their English translations (figure 3.6):[35]

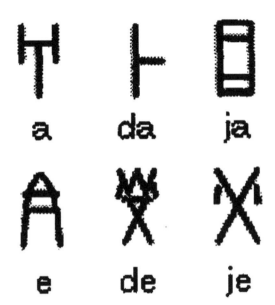

a   da   ja

e   de   je

FIGURE 3.6
Linear B Syllabic Signs.
Source: **http://www.omniglot
.com/writing/linearb.htm**

Chadwick adds, "This elementary deduction was neglected by many of the would-be decipherers."[36] It is a deduction based on their empirical knowledge of other languages. These are among the premises of their argument:

A. If a script is ideographic, then it must have many hundreds (or thousands) of signs.
B. If a script is alphabetic, then it cannot have eighty-nine separate alphabetical characters.
C. Linear B has more than eighty-nine and fewer than 160 signs.

Neither *A* nor *B* proposition is logically required. We can imagine a society that had only a few dozen signs in its written script, and we can imagine a society whose language had eighty-nine separate alphabetical characters.

However, it appears that no known human language has so few logograms, and none has more than forty distinct alphabetical characters.[37] Assuming that they had a large enough sample of Linear B, Chadwick and Ventris's argument is valid: Linear B must be syllabic, not alphabetical.[38]

New Insight into Old Wine: Fermentation Studies in 1847

In 1720 Newton listed fermentation among those processes not understood; it was a black box whose inner workings were occult. This changed with the epochal works of Louis Pasteur (1822–1895) beginning around 1847 when he studied molecular asymmetry. For this work Pasteur depended on new discoveries and experimental techniques from crystallography, chemistry, and optics. Based on these new discoveries, made possible by using improved microscopes to see into the black box of otherwise invisible life forms, Pasteur moved fermentation out of the realm of occult events and into the realm of scientific understanding. Thus, he could say, "Ferments, properly so called, are living beings, that the germs of microscopic organisms abound in the surface of all objects, in the air and in water."[39]

Pasteur's "On the Relations Existing between Oxygen and Yeast" (1879) offers a classical description of the type-one black box. Having demonstrated some ten years earlier that living entities cause fermentation, Pasteur notes that it remains to be seen how ferments work and whether or not these newly discovered processes are explicable "by the ordinary laws of chemistry."[40] In other words, do fermentation processes obey or violate the laws of chemical interaction which do not admit of occult properties?

In this 1879 paper Pasteur refers back to earlier research that, he says, has gained in value because it contained predictions subsequently borne out: "It is a great presumption in favor of the truth of theoretical ideas when the results of experiments undertaken on the strength of those ideas are confirmed by various facts more recently added to science and when those ideas force themselves more and more on our minds, in spite of a prima facie improbability." He summarizes key aspects of a paper, from June 28, 1861, titled "Influences of Oxygen on the Development of Yeast and on Alcoholic Fermentation."[41]

Pasteur reduces formerly mysterious biological events to observable biochemical events. We can summarize his claims:

1. General observations, "Alcoholic ferments must possess the faculty of vegetating and performing their functions out of contact with air."
2. Thus, the essential processes are *anaerobic*, a term that Pasteur coined.

3. "If we compare under the microscope the appearance and character of the successive quantities of yeast taken, we shall see plainly that the structure of the cells undergoes a progressive change."

4. "Ordinary moulds assume the character of a ferment when compelled to live without air or with quantities of air too scant to permit of their organs having around them as much of that element as is necessary for their life as aerobian plants."

5. "These results prove clearly that the fermentative character is not an invariable phenomenon of yeast life; they show that yeast is a plant that does not differ from ordinary plants and that manifests its fermentative power solely in consequence of particular conditions under which it is compelled to live."

6. This predicts the discovery that fermentation occurs in other cells.[42]

Paul Decelles notes that there are additional fermentation pathways: "Yeast cells produce ethyl alcohol by fermentation. Certain cells of our body, namely muscle cells, use lactic acid fermentation, while depending on the organism some of the other products of fermentation include acetic acid, formic acid, acetone, and isopropyl alcohol."[43]

Another, less-imposing, puzzle appears when we reconstruct conscious intentions that we fail to carry out. Sometimes I discover, for example, that I did not secure a cup of coffee at Alpine Bagel on 21st Avenue although that was my intention. Without too much effort, I can specify a typical set of inputs: I am coffee dependent; I always intend to get coffee at Alpine; I set aside two one-dollar bills. And I can specify and measure the output: I usually have in my hands a large French roast (with extra half-and-half) as I cross 21st Avenue. Sometimes, I end up not getting the coffee. In these circumstances, I can reconstruct what happened by recounting the usual inputs and the aberrant output. On one occasion just before I entered Alpine Bagel I saw David, we chatted for fifteen minutes, I heard the bell chime, and I raced to my first class, sans French roast.

Specifying Inputs and Outputs, Specifying If/Then Statements, and Experiments

I intentionally throw into this first box three different phenomena. I suggest that ordinary science of the kind that Ashby focuses on does not depend on the subject matter—machines, neural tissue, mental contents—but, rather, on the protocols we can discover through observing them over time.[44]

To put this in logical terms, specifying inputs is identical to specifying *if* statements; specifying outputs is identical to specifying *then* statements. More so, *if* statements denote what are typically called *initial conditions* in an experiment; *then* statements denote what are typically called *outcomes* in an experiment; their conjunction specifies a low-level theory. This may seem trivial, but it is not. All the vastly complex calculations carried out by classical scientists (and their predecessors), like those done on today's fastest computers, depend on if/then reasoning of precisely this kind.

To go back to my mundane example, if I intended to get a French roast, if I did not get the French Roast, and if everything else was as usual, *then* something interfered with my intentions. Looking backward, I realize that my intense chat with David interfered with my conscious intentions to secure a steaming cup of French roast.

Following a careful series of recordings, we arrange our observations into *if* statements and *then* statements, and, by joining them, form causal propositions. Thanks to the geniuses of nineteenth-century mathematics and early twentieth-century information theory, we know that *if* and *then* statements (and all their cousins) can be mapped directly into electronic form via circuits that mimic logical formulations. According to Charles Petzold, we can thank Claude Shannon for showing in his 1938 MIT master's thesis that Boolean expressions could be expressed in electrical circuits: "Nobody has shown with Shannon's clarity and rigor that electrical engineers could use all the tools of Boolean algebra to design circuits with switches."[45]

Type-Two Black Boxes: Educated Hunches

We know that the Enigma machine worked: engineers designed it to encode natural-language messages. German soldiers and sailors used improved Enigma machines on a daily basis to send and receive messages. Given sufficient time, sufficient cleverness, and immense labor, English and Polish scientists revealed most of Enigma's secrets. In the same way, when in an attempt to avoid paying royalties a Chinese company seeks to duplicate an American gizmo, its engineers know that they can treat the gizmo as a type-one black box and—with enough resources—reverse engineer it. The economic costs of reverse engineering may be too great, but the conceptual task is clear: figure out what each part of the thing does, and then figure out how it works with all the other parts.

With type-two black boxes, this is not such a simple proposition (figure 3.7):

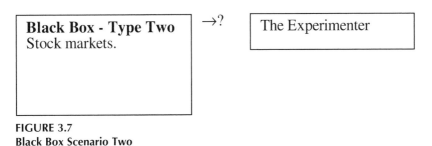

FIGURE 3.7
Black Box Scenario Two

In this scenario we cannot *easily* specify inputs and cannot *easily* specify outputs. There may be many reasons for this limitation. One reason is that we may not know what counts as an input, or we may not know how to isolate one kind of input from another. In a similar way, we often do not know how to specify what counts as an output or how to discern one kind of output from another. This means that we cannot easily specify *if* statements and that we cannot easily specify *then* statements. This has two unhappy consequences: One is that we cannot use if/then reasoning, which means that we cannot use ordinary rules of logic. We cannot, in that sense, hope to use deductive reasoning the way that Chadwick and Ventris did when they deduced that Linear B must be a syllabic script. A second unhappy consequence is that, lacking the ability to specify inputs and outputs, we cannot create what Ashby called *the protocol*, and thus we cannot create a canonical representation.

Because we can only guess at inputs and outputs, we cannot formulate representations that are demonstratively isomorphic with the object under investigation. This means that we cannot offer models of the black box that will be persuasive to all reasonable persons. Because we have had to guess at what count as inputs and guess at what count as outputs, we cannot defend our choices on logical or empirical grounds. (If we could, we'd then have a type-one black box.) Lacking logical and empirical grounds, we are left with hunches, biases, intuition, et cetera—each of them a source of ideas, none of them clear and distinct.

We might put numerous examples into this black box—most of what counts as psychotherapy and theories of education, for example, fall into this category. Especially intuitive or persuasive people can offer plausible accounts of what should count as inputs and outputs. We see this in accounts of stock markets and similar arenas where special, intuitive talents seem required to grasp what should count as evidence for focusing on one set of inputs versus another. George Soros, a renowned hedge-fund manager and financial wizard, wrote a lengthy book about how he thought through market conditions.

Soros makes this point repeatedly in his many essays on what he terms *reflexivity*. This describes the effect that beliefs about the future have on complex events, from individual actions to gigantic institutions. By seeking to understand stock markets, and acting on those beliefs, we alter the thing we wished to predict: "On the one hand, participants seek to understand reality; on the other, they seek to bring about a desired outcome. The two functions work in opposite directions: in the cognitive function, reality is the given; in the participating function, the participants' understanding is the constant."[46] Thus, even in those circumstances where we can designate likely inputs—like today's Dow Jones Industrial Average (DJIA)—we cannot predict precisely its future behavior unless we can also discern how *other* market participants understand the DJIA at this moment. More so, if we share our opinions—say, for example, that the market is oversold—that diagnosis becomes another, causal, agent in the nexus of events that shape the market.

As Soros notes, "Theories about human behavior can and do influence human behavior."[47] The so-called Freud wars of the 1990s in the United States centered on the effect that Freudian and related social-science theories had on children, especially on their religious and sexual beliefs and the validity of repressed memories.[48]

Type-Three Black Boxes: Plausible Guessing

Type-three black boxes do not permit us to consistently specify any inputs or outputs. We are left, therefore, with guesswork. Guesswork may be inspired or bizarre; lacking the ability to evaluate one guess against another, we are left with persuasion, brilliance, authority, and other nonscientific measures that may or may not be valid. Newton listed fermentation as an occult event; it was, in our terms, located in a type-three black box. I believe most of psychotherapy remains locked in the same kind of box (figure 3.8).

The neurosciences potentially offer students of religion what they potentially offer students of psychotherapy—a way to winnow out good theories

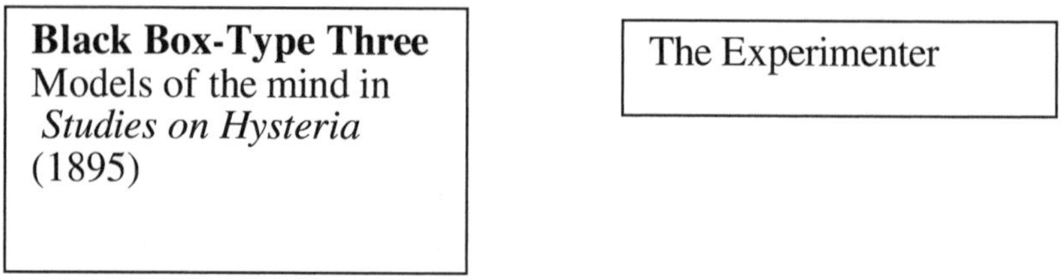

Black Box-Type Three Models of the mind in *Studies on Hysteria* (1895)	The Experimenter

FIGURE 3.8
Black Box Scenario Three

from bad. Using Ashby's terms, we want to move psychotherapy and religious studies from their residence in type-three black boxes uptown into type-two black boxes.

When Freud helped create psychotherapy around 1895, he had three tasks before him: to exploit the neurosciences of his day, to distinguish psychotherapy from pastoral counseling, and to extract psychoanalytic theory from philosophy. Numerous commentators have attacked or defended Freud's success in these three ventures. I sidestep these debates and instead use the model of the black box to recast Freud's efforts to ground psychoanalytic psychotherapy. I show that, despite his best efforts, Freud was not able to specify inputs and outputs in ways that match the criteria of type-one black boxes. One reason for this failure was his philosophic commitments. Although trained in laboratory neuroscience, Freud's metaphysics required him to disdain "mere appearances" in favor of conjectures about unseen processes.[49] This led him to create a series of brilliant narratives about those unseen mental processes that his philosophy required him to assert were otherwise invisible.[50]

A World Shorn of God and Shorn of Qualities

Freud says that reality testing occurs between an out-there real world and an in-here real perceiver. His philosophical commitment to materialism leads him to locate qualitative experience within the mind alone and to see human beings as experiments of nature: "We will still show too little respect for Nature, which (in the obscure words of Leonardo, which recall Hamlet's lines) 'is full of countless causes [*ragioni*] that never enter experience.' Every one of us human beings corresponds to one of the countless experiments in which these *ragioni* of nature force their way into experience."[51]

The world is shorn of God,[52] and it is shorn of quality because science teaches us that the real world consists only of varying amounts and concentrations of quantities.[53] This fundamental reality is prior to humans and hence must be that which created us. Our scientific task is to correct the errors of our faulty education and faulty libidinal histories, especially our group narcissism, in order to discover that underlying, unconstructed reality. Knowing the truth sets us free not because reason is more powerful than the passions but because it is the surer path to knowledge. We are not in danger of destroying the unconstructed world of reality by attacking the worlds of meaning (*Sinn*) because they are competing attempts to understand the real world. The anxieties and fears that beset religious persons when the sacred is challenged are born out of their unwillingness to abandon pleasurable fantasies, not out of insight that reality itself is endangered.

Claude Lévi-Strauss, the maestro of structural anthropology, says that intellectuals of his generation believed that Marxism and psychoanalysis uncovered latent forces that shaped their personal and collective histories. To these discourses Lévi-Strauss added geology as a sister discipline: "All three showed that understanding consists in the reduction of one type of reality to another; that true reality is never the most obvious of realities, and that its nature is already apparent in the care that it takes to evade our detection."[54]

Using Ashby's terms, this seems exactly wrong. Geological investigations typically permit one to designate inputs and outputs; Freudian and Marxist critiques rarely (if ever) show this kind of clarity. For that reason, most Freudian and Marxist efforts occur in a type-three black box. While all three disciplines examine the past, geology alone offers ways to systematically alter inputs and record the effect on outputs. That means that geologists alone employ direct, evidence-based ways to support some hypotheses and reject others. When luck runs their way, they can offer models of a geological event to the level of an isomorphism.

Just as Ventris was able to conduct scientific experiment on Linear B, geologists can evaluate conjectures about geological events, even if they occurred a long time ago. Freud's effort to discover the hidden forces that control human personality failed (and succeeded) in ways analogous to David Hume's efforts to create a new scientific philosophy in his masterpiece *A Treatise of Human Nature* (1739–1740).[55] We can compare Hume's book with that of his near-contemporary, the geologist James Hutton. Hume and Hutton were intellectual compatriots, equally gifted and equally vigorous. Yet Hutton's *System of the Earth* (1785) entered into ordinary science, while Hume's work did not.[56] Hume's failure to establish a foundation for the construction of the moral sciences was a noble one. His task—the creation of a rational theory of human nature—remains unfinished two and a half centuries later. Hutton's conjectures have been winnowed through subsequent observations to the degree that students of geology need never read his 1785 book.

One reason Hume and Freud did not secure a scientific foundation is that neither offered ways to link his conjectures to directly observable events, much less to events that could be manipulated systematically. Hume struggled to define morality, for example, using the concept of taste. While this is suggestive, it has not led to increased clarity. Indeed, common to Humean and Freudian studies is a struggle to define exactly what Hume or Freud *meant* by their key concepts.[57] A similar task falls to those who seek to clarify exactly what Max Weber meant in his seminal text, *The Protestant Ethic and the Spirit of Capitalism* (originally 1905). In their new translation, Peter Baehr and Gordon C. Wells note that in order "to join the pantheon of literary works that are commonly called classics, Weber's essay also had to possess qualities of fruitful ambiguity, or textual suppleness, which allow multiple readings and adoptions."[58]

Weber's ambiguous language, his tropes, and his reliance on the concept of ideal types render his arguments irrefutable. He and his critics need never agree on the meaning of these key concepts; hence, while his essay provoked numerous responses, and even polemics, it is not capable of yielding a singular reading.

Lacking that clarity, schools of thought emerge, each defending a reading of Hume, Freud, or Weber hotly contended by members of a second or third school. For example, in Hume's theory of ethics taste denotes a special kind of perception; in Kant's theory of aesthetics taste denotes a synthetic a priori judgment that allows one to ascertain the beautiful.[59]

Like Hume, Freud sought a reasonable balance between competing and antithetical views of human nature, particularly sentimentalism and intellectualism.[60] Like Marx, Freud used political and interpersonal concepts to explain unseen mental processes. The psychological concept of resistance is central to Freudian and Marxist explanations of human behavior. By emphasizing the strength with which nature, including human nature, resists conceptualization, Lévi-Strauss aligns himself with Freud's and Marx's philosophies of science and philosophies of religion. All three affirm that superficial structures and self-understandings of a society are masks that distort a hidden reality that superior reasoning and superior courage will someday uncover. By insisting that the irrational features of group and individual behaviors are the results of hidden but intentional forces, Freud and Marx illustrate what Paul Ricoeur called *the hermeneutics of suspicion*. Both explain suffering as the product of secret intentions; Freud exposed the part of the person that secretly controls sexuality and aggression, whereas Marx exposed the part of society that controls the modes of production and masks its crimes with religious coverings.

Originating as critiques of manifest content, both hermeneutics require us to challenge our naïve self-understanding. Not accidentally—because Marx and Freud offered complete theories of human being, including its hiddenness—both declared themselves enemies of religion. Freudianism and Marxism emerged as contending schools of theological anthropology: if they were correct, theism—certainly Jewish and Christian narratives—must be wrong. The would-be analyst and the would-be Marxist critic cannot enter into the realms of critical theory without first examining their unanalyzed consciousness. To use terms from Ashby, authorities in both systems assume that persons with unanalyzed, naïve consciousness cannot discern real inputs and real outputs of the system in question. Lost in errors, naïve critics cannot grasp the disguised wishes of the neurotic or the hidden motives of the bourgeois. This leads theoreticians in both schools to challenge the relevance of criticisms that do not arise out of an examination of one's personal and socially repressed unconscious. At their extreme, both schools demand that critics be born again, their consciousness raised before their criticisms merit hearing.

Speculations about the Inner Mind: *Studies on Hysteria* as Type-Three Black Box

In *Studies on Hysteria* (1895), coauthored with the renowned Viennese physician Josef Breuer, Freud offered his first published model of the inner world, the unconscious mind. In the 1880s Breuer had treated the famous patient known as Anna O. and had subsequently invited Freud to coauthor a series of papers, which eventually culminated in their jointly authored book. Breuer argued that Anna's hysteric symptoms, such as nervous coughing and the like, were formed in hypnogogic states.[61] He used hypnosis to induce and then rectify those original hypnogogic traumata.

Breuer claimed that hysteric symptoms were repetitions of past, but forgotten, traumata whose associated affect could be drained by analyzing the origins of subsequent symptoms and so restore the patient's memory to its unrepressed state. Breuer discovered that he and Anna inevitably ended up tracing back, in sequence, the entire train of events connected to a particular symptom. For example, Breuer reported that he had to trace 108 separate memories related to not hearing a person enter the room before Anna gained partial relief from her hysterical deafness.[62] She gained total relief only after Breuer had traced back the memories of 195 other events related to not hearing someone under different circumstances.[63] Given this apparently justified deterministic principle, that consciousness proceeds to regain its lost territories by sequential steps, it follows, as Freud put it, that "every single reminiscence that emerges during an analysis of this kind has significance. An intrusion of irrelevant anemic images . . . in fact never occurs."[64]

Freud reported that he found hypnosis useless: "Strangely enough, I have never in my own experience met with a genuine hypnoid hysteria. Any that I took into my hands has turned into a defense hysteria."[65] Freud describes how he came to believe that hysterical symptoms represent psychological conflicts between intrapsychic forces. Most of Freud's patients could not recall much about their symptoms. When Freud used hypnosis to induce better recall, he found that they frequently did not recognize the "memories" they had reproduced while in the hypnotic trance. So Freud changed his technique:

> I now became insistent—if I assured them that they did know it, that it would appear to their minds—then, in the first cases, something did actually occur to them. . . . After this I became still more insistent; I told the patients to lie down

and deliberately close their eyes in order to "concentrate." . . . I then found that without any hypnosis new recollections emerged that went further back and that probably related to our topic. Experiences like this made me think that it would in fact be possible for the pathogenic groups of ideas, that were, after all, certainly present, to be brought to light by mere insistence; and since this insistence involved effort on my part and so suggested the idea that I had to overcome a resistance, the situation led me at once to the theory that *by means of my psychical work I had to overcome a psychical force in the patients that was opposed to the pathogenic ideas becoming conscious (being remembered).*[66]

In other words, Freud projected the interpersonal struggle between himself and his patient into his patient's mind. By mapping the two-person conflict into his patient's mind he can explain the patient's resistance against getting well: "A new understanding seemed to open before my eyes when it occurred to me that this must, no doubt, be the same psychical force that had played a part in the generating of the hysterical symptom and had at that time prevented the pathogenic idea from becoming conscious."[67]

Freud couched his claims in the physicalist language of the so-called Helmholtz school.[68] In his letters to Fliess, we see him try out and abandon literally dozens of plausible hypotheses on the origins of hysteric symptoms.[69] However, this model becomes less plausible when Freud states that many patients *never* recall the crucial scenes that they and their doctor laboriously reconstructed:

Sometimes it [pressure on the forehead] calls up and arranges recollections that have been withdrawn from association for many years but that can still be recognized as recollections, and sometimes, finally, as the climax of its achievements in the way of reproductive thinking, it causes thoughts to emerge that the patient will never recognize as his own, that he never *remembers*, although he admits that the context calls for them inexorably, and while he becomes convinced that it is precisely these ideas that are leading to the conclusions of the analysis and the removal of his symptoms.[70]

When Freud says that the context calls for such memories to be there—even in the absence of an actual memory—we have a statement of faith, not an observation. It demonstrates his materialist metaphysics. That metaphysical commitment, not direct observations of the inputs and outputs, provided the context into which the never-recovered memories must fit.

And, yet, Freud reports that patients got well once they became convinced of the accuracy of the reconstruction. In 1895, Freud explained this by referring to the abreaction of pent-up psychic energies that had been misdirected because of early traumata, whose binding produced psychic pain and whose liberation gave psychic health. Freud emphasizes that his kind of treatment requires commitment much greater than that required for patients who are physically ill:

> The procedure is laborious and time-consuming for the physician. It presupposes great interest in psychological happenings, but personal concern for the patients as well. I cannot imagine bringing myself to delve into the psychical mechanism of a hysteria in anyone who struck me as low-minded and repellent and who, on closer acquaintance, would not be capable of arousing human sympathy; whereas I can keep the treatment of a tabetic [wasting, especially with syphilis] or rheumatic patient apart from personal approval of this kind.[71]

Treatment is hard, unpleasant, tedious work that requires the doctor to constantly invigilate against his patient's conscious and unconscious attempts to relapse, to resist insight, to block the reconstruction of archaic events, and, as it were, to avoid a full confession.[72]

While he locates the scene of such conflicts and confrontations *within* his patients' psyches, Freud is frank enough to note that patient and doctor must be emotionally committed to one other.[73] It is this real, interpersonal, relationship that provides the fulcrum by which the physician persuades his patient to abandon her essentially private role of neurotic sufferer. Thus Freud concludes by addressing his hypothetical patient: "You will be able to convince yourself that much will be gained if we succeed in transforming your hysterical misery into common unhappiness. With a mental life that has been restored to health you will be better armed against that unhappiness."[74]

Evident in Freud's eloquent prose are his commitment to the ideal of material causation, his sympathy for his patient, his patient's desire for cure, and the strength of the relationship that quickly developed between them. In 1895 Freud overlooks those factors and chooses to explain his patient's behaviors by referring back to and reconstructing their mysterious origins. (This changes in his later texts.)

Freud's creativity and importance are unquestionable; he helped shape the Western mind, and we are his heirs. Yet, equally unquestionable are many as-yet-unsolved problems with psychoanalytic theory. We can explain some of those problems by referring back to Ashby's description of what I have called a type-one black box. When Freud explained his patients' external conflicts by locating them within an unseen part of the patients' mind (in the unconscious), he sought to explain the workings of the black box without first knowing what counted as inputs and what counted as outputs. He rightly feared that critics would reduce psychoanalysis to yet another type of persuasion. Persuasion and spiritual healing have, in fact, occurred for millennia in all human societies. The most rigorous studies of outcome measures of successful psychotherapies affirm that (1) therapy works well, perhaps 80 percent of the time, and that (2) we don't know how it works.

In *The Great Psychotherapy Debate: Models, Methods and Findings* (2001),[75] Bruce Wampold reviews outcome studies of the effectiveness of various forms of medical psychotherapy and, thus, the validity of their underlying theories of pathogenesis. Adherents of the medical model presume that, having identified causal agents, the psychotherapist uses precise therapeutic agents to counter them and to cure the patient's psychological disease. When Freud guessed that hysterical patients suffered mainly from their memories of partly abreacted traumata, his guess led him to emphasize recovering memories by dredging up these memories, thus achieving a cure. When contemporary adherents of cognitive-behavioral therapy guess that many forms of depression, for example, derive from punitive self-talk, they focus their therapeutic efforts on confronting these internal dialogues and thus effect cure.

In each school, adherents focus on specific causal factors—what Wampold calls "specific ingredients"—as the agent most responsible for cure. Because psychotherapy, as an aggregate, has an effectiveness rating of nearly 80 percent, these are significant claims. On the one hand, evidence is strong that psychotherapy works well. On the other hand, that there is little or no evidence for the medical model. Wampold asserts, "If the medical model provides a useful framework for conceptualizing psychotherapy, then evidence should suggest that the specific ingredients are responsible for the benefits of psychotherapy."[76] Summing up his review of outcome studies, Wampold says that the combined evidence for specific ingredients hovers between 0 percent and 1 percent: "Decades of psychotherapy research have failed to find a scintilla of evidence that any specific ingredient is necessary for therapeutic change."[77]

If we agree that there are no specific ingredients and that therapy works, then there must be general elements—what Wampold calls contextual elements—that distinguish effective psychotherapy from ineffective psychotherapy. Here he joins ranks with Jerome Frank, who made this point some forty-five years ago in *Persuasion and Healing*.[78] Wampold says that these common factors are

(a) an emotionally charged confiding relationship with a helper; (b) a setting that evokes expectation of help; (c) a rationale, conceptual scheme, or myth that provides a plausible, although not necessarily true, explanation of the client's symptoms and how the client can overcome his or her demoralization; and (d) a ritual or procedure that requires the active participation of both client and therapist and is based on the rationale underlying the therapy.[79]

Not accidentally, Wampold uses the terms *ritual* and *myth*, which derive from religious studies and the anthropology of religion. Indeed, Claude Lévi-Strauss made this same point in a famous essay in which he compared New York psychoanalysts to ancient shamans.[80]

If we cannot designate what counts as inputs, we cannot alter them systematically. This means that we cannot establish protocols and thus cannot establish canonical representations. Lacking canonical representations, we cannot hope to offer isomorphic models of the workings of the unseen interior. Instead, schools of thought spring up, each battling the other, each led by sometimes-impassioned leaders who thunder across the valleys that separate one camp from another.[81]

We would expect to find that such successful modes of identity transformations take place in highly ritualized contexts. And of course this is true of any long-term psychotherapy. It is especially true of the early psychoanalytic movement when the dangers of extreme heterodoxy, hostility, ignorance, and resistance on the part of official science compounded Freud's premonition that his truths would be resisted and, if possible, destroyed.[82]

To say that psychoanalytic psychotherapy is a complex *rites de passage* similar to religious rituals is not to libel it.[83] On the contrary, psychoanalysis makes use of and exhibits the wisdom of personal transformation known to all successful religious psychotherapies. This conclusion contradicts Freud's wishes to divorce his creation from religion and to ally it with the sciences. The history of psychoanalysis shows a number of brilliant persons trying to make this happen, without success. We know that gifted therapists can help transform human lives. We know, existentially, that human beings cannot live without a sense of being loved and that ancient religious truths and modern psychotherapies typically affirm the validity of one another. Grounding those truths in neuroscience is a task that Freud began in the nineteenth century and that we may yet see completed in the twenty-first.

Notes

1. See J. J. O'Connor and E. F. Robertson, "René Descartes," at http://www-gap .dcs.st-and.ac.uk/~history/Biographies/Descartes.html (accessed November 14, 2008). Descartes "makes the first step toward a theory of invariants, which at later stages derelativises the system of reference and removes arbitrariness."

2. W. Ross Ashby, *An Introduction to Cybernetics* (London: Chapman & Hall, 1956). Also available free in PDF form online at http://pespmc1.vub.ac.be/books/ IntroCyb.pdf (accessed November 14, 2008).

A polemical use of the term *black box* appears in a notorious attack on Darwinian thought in Michael J. Behe, *Darwin's Black Box: The Biochemical Challenge to Evolution* (New York: Simon & Schuster, 2004).

3. Ashby, *Introduction to Cybernetics*, 86.

4. Ashby, *Introduction to Cybernetics*, 87.

5. Karl L. Wuensch, "When Does Correlation Imply Causation?" at http:// 72.14.235.104/search?q=cache:PAUfJBK8lg8J:core.ecu.edu/psyc/wuenschk/StatHelp/

Correlation-Causation.htm+%22correlation+does+imply+causation%22&hl=en&g
l=au&ct=clnk&cd=1&client=firefox-a (accessed November 14, 2008). Wuensch, of
East Carolina University, says, "When the data have been gathered by experimental
means and confounds have been eliminated, correlation does imply causation."

6. See Stevan Harnad, "Correlation vs. Causality: How/Why the Mind/Body
Problem Is Hard," *Journal of Consciousness Studies* 7, no. 4 (2000): 54–61, and online
at Cogprints, http://cogprints.org/1617/ (accessed November 14, 2008). For example,
if "mind can cause brain events"—assuming that is a meaningful phrase—then we
assume that the mental event preceded the brain event. See also Christof Koch, *The
Quest for Consciousness: A Neurobiological Approach* (Englewood, Colo.: Roberts and
Company, 2004), 18–19.

7. Ashby, *Introduction to Cybernetics*, 88.

8. Ashby, *Introduction to Cybernetics*, 90.

9. Ashby, *Introduction to Cybernetics*, 93.

10. Ashby, *Introduction to Cybernetics*, 94.

11. Michael S. Mahoney, "The Mathematical Realm of Nature," at http://www
.princeton.edu/~mike/articles/mathnat/mathnatfr.html (accessed November 14,
2008). Mahoney notes, "Converging in the concepts and techniques of infinitesimal
analysis, rational mechanics [in the seventeenth century] became a branch of mathematics, and mathematics opened itself to mechanical ideas. The symbolic algebra and
the theory of equations from which infinitesimal analysis took inspiration and form
were aimed initially at abstracting mathematics from the concrete world and had the
effect of freeing it to create imaginary and counterfactual structures irrespective of
their real or even possible instantiation."

12. C. B. Boyer, *The Concepts of the Calculus* (1959), 21.

13. Aristotle, *Metaphysics*, trans. Stephen Makin (Oxford: Clarendon Press, 2006),
1001b; Boyer, *Concepts of the Calculus*, 23.

14. Volney Gay, *Progress in the Humanities?* (forthcoming).

15. All three citations of *Principia* are from Dana Densmore, *Newton's* Principia:
The Central Argument (Santa Fe, N.M.: Green Lion Press, 1995), 2.

16. Newton in Densmore, *Newton's* Prinicipia, 405.

17. Newton in Densmore, *Newton's* Prinicipia, 406.

18. See I. Bernard Cohen, "*The First English Version of Newton's Hypotheses non
fingo*," *Isis* 53, no. 3 (1962): 379–88. See also Andrew Janiak, "Newton's Philosophy,"
October 13, 2006, at http://plato.stanford.edu/entries/newton-philosophy/ (accessed
December 10, 2009). We know that Newton expended prodigious energy in other
contexts precisely on these questions: "Newton's discussion in *De Gravitatione* is also
related to, and helps to illuminate, the more famous discussion of space and time in
the General Scholium to the *Principia*. In each of these texts, Newton apparently transcends the narrow questions concerning space, time, and motion raised by physic's
need to distinguish between true and apparent motion, tackling broader questions
whose origin seems to lie in the seventeenth-century metaphysical tradition."

19. Cited by Cohen, "Newton's *Hypotheses*," 386n23. Texts taken from Isaac Newton, *Opticks*, ed. I. B. Cohen and Duane H. D. Roller (New York: Dover, 1952), 357.

20. Newton, *Opticks*, 376.

21. Newton, *Opticks*, 401.

22. Densmore, *Newton's* Principia, xxixn7.

23. See Principia Cybernetica Web, "Black Box Method," at http://pespmc1.vub
.ac.be/asc/black_metho.html (accessed November 14, 2008). As Klaus Krippendorff
put it, "The isomorphism between the black box and its model, which the method
aims to establish, does not imply structural correspondences between the two. One
may be a mechanical device, a chemical process, or human organ, the other may be
a mathematical formula, an algorithm, or electronic piece of equipment. However,
the organization of such a model often leads to fruitful hypotheses concerning the
structure of the black box."

24. Gay, *Progress in the Humanities?*

25. See an image of an Enigma machine at http://www.mccullagh.org/db9/10d-13/
enigma-machine.jpg (accessed November 14, 2008).

26. Wikipedia.org, "bombe," at http://en.wikipedia.org/wiki/Bombe (accessed
November 14, 2008). "In the bombe, a set of rotors with the same internal wiring as
the German Enigma rotors was used but designed to be spun by a motor, stepping
through all possible rotor settings. The bombe rotors had a double set of contacts and
wiring to emulate the Enigma reflection."

27. See Władysław Kozaczuk, *Enigma: How the German Machine Cipher Was Broken,
and How It Was Read by the Allies in World War Two*, ed. and trans. Christopher Kas-
parek (Frederick, Md.: University Publications of America, 1984). Its name and proto-
type derive from the work of Polish mathematicians, especially Marian Rejewski.

28. See "Turing's Treatise on Enigma (The Prof's Book)," at http://cryptocellar
.org/Turing/ (accessed November 14, 2008). Early and simpler versions of the machine appeared commercially in the 1920s. Some ciphers done with more sophisticated German Navy versions of Enigma remained unsolved until 2006. Turing wrote
about his work at Bletchley Park.

29. John Chadwick, *The Decipherment of Linear B* (Cambridge: Cambridge University Press, 1958). For the technical arguments, see Michael Ventris and John Chadwick,
Documents in Mycenaean Greek (Cambridge: Cambridge University Press, 1956).

30. Chadwick, *Decipherment*, 67.

31. See Dan Sperber, *Rethinking Symbolism* (Cambridge: Cambridge University
Press, 1974). That mythic systems and many forms of religious belief *affirm* contradictions as central to their core beliefs is an important fact about religion.

32. "Bombe," Wikipedia.org.

33. Ventris and Chadwick, *Documents*, 43.

34. Both sets come from Simon Ager, "Linear B," in *Omniglot: Writing Systems and
Languages of the World*, at http://www.omniglot.com/writing/linearb.htm (accessed
November 14, 2008).

35. Chadwick, *Decipherment*, 43.

36. See Simon Ager, "Georgian," in *Omniglot*. Some claim that the largest alphabet
is Georgian, first written in the seventeenth century. It has about forty characters.

37. They use *Modus tollens* in their deduction: If P, then Q; not-Q, therefore not-P.
In support of this deduction we know that either Q or not-Q. Assume if P, then Q; if
not-Q, then P. This cannot be valid, because we assumed P, then Q; if we deduce P,
then Q must follow.

38. Originally read before the French Academy of Sciences, April 29, 1878. Published in *Comptes Rendus de l'Académie des Sciences* 86:1037–43. Reprinted and translated as "Scientific Papers," in *The Harvard Classics*, vol. 38, no. 7 (New York: P. F. Collier & Son, 1909–1914). Available online at http://www.bartleby.com/38/7/ (accessed November 14, 2008).

39. Modern History Notebook, "Louis Pasteur (1822–1895): Physiological Theory of Fermentation," trans. F. Faulkner and D. C. Robb, at http://www.fordham.edu/halsall/mod/1879pasteur-ferment.html (accessed November 14, 2008).

40. See Bartleby.com, "Louis Pasteur (1822–95): Scientific Papers; The Harvard Classics, 1909–14," at http://www.bartleby.com/38/7/1.html (accessed November 14, 2008). The passage is from the *Bulletin de la Société Chimique de Paris*.

41. Bartleby.com, "Louis Pasteur." Pasteur died before the first Nobel Prizes were awarded in 1901, but the study of fermentation led to at least one prize: the 1907 prize for chemistry was awarded to Eduard Buchner "for his biochemical researches and his discovery of cell-free fermentation." (Nobelprize.org, "The Nobel Prize in Chemistry 1907," http://nobelprize.org/nobel_prizes/chemistry/laureates/1907/ [accessed June 10, 2009]).

42. Paul Decelles, "Fermentation and Anaerobic Respiration," at http://staff.jccc.net/pdecell/cellresp/fermentation.html (accessed November 14, 2008).

43. George Kampis, "Explicit Epistemology," at http://hps.elte.hu/~gk/Publications/JapanEE.pdf (accessed November 14, 2008). "A form of black boxing plays a central role in science. Consider how we obtain our knowledge in the laboratory. In comes the experimental question, out goes the empirical result. The magic in between is done by Nature behind the veil, so we don't know it in the first place. How the magic is done we learn by forming theories to fill the gaps between the two ends."

44. Charles Petzold, *Code: The Hidden Language of Computer Hardware and Software* (Redmond, Wash.: Microsoft Press, 2000), 103. Claude Shannon's thesis appeared in "A Symbolic Analysis of Relay and Switching Circuits," *Transactions American Institute of Electrical Engineers* 57 (1938): 713–23.

45. George Soros, *The Alchemy of Finance* (New York: John Wiley & Sons, 1987), 2.

46. Soros, *Alchemy*, 7.

47. See, for example, Peter Gay, *Pleasure Wars—the Bourgeois Experience: Victoria to Freud* (New York: Norton, 1993); Harold P. Blum et al., *The Memory Wars: Freud's Legacy in Dispute* (New York: New York Review of Books, 1995); John Forrester, *Dispatches from the Freud Wars: Psychoanalysis and Its Passions* (Cambridge, Mass.: Harvard University Press, 1997); Stephen Wilson, *Introducing the Freud Wars* (Palmerston North, New Zealand: Totem Books, 2003); and Lavinia Gomez, *The Freud Wars: An Introduction to the Philosophy of Psychoanalysis* (New York: Routledge, 2005).

48. Volney Gay, *Freud on Sublimation* (Albany: State University of New York, 1992). In my book I evaluate the literature on Freud's epistemology and explore the issue of quality and quantity.

49. Cornelius Borck, "Visualizing Nerve Cells and Psychical Mechanisms: The Rhetoric of Freud's Illustrations," in *From Brain Research to the Unconscious*, ed. Giselher Guttmann and Inge Scholz-Strasser (Vienna: Verlag der Österreichischen Akademie der Wissenschaften, 1998), 57–86, 60. Freud used gold chloride to study nervous tissue and made exact renderings. Cornelius Borck evaluated Freud's drawings

and notes, saying, "What strikes us once again are the precise details of the illustration" (63, following). This exactness disappears in Freud's psychoanalytic diagrams, none of which offers isometric renditions. I cite Borck in my forthcoming book, Gay, *Progress in the Humanities?*

50. Freud, *Leonardo da Vinci and a Memory of His Childhood*, vol. 11, *The Standard Edition of the Complete Psychological Works of Sigmund Freud*, ed. J. Strachey (London: Hogarth Press, 1957), 137.

51. Paul Ricoeur, *Freud and Philosophy* (New Haven, Conn.: Yale, 1970), 327.

52. Sigmund Freud, *The Origins of Psycho-analysis: Letters to Wilhelm Fliess, Drafts and Notes, 1887–1902*, eds. Marie Bonaparte, Anna Freud, and Ernst Kris; trans. Eric Mosbacher and James Strachey; intro. Ernst Kris (Imago: London and New York, 1954). Partly including "A Project for a Scientific Psychology," *Standard Edition* 1:175. The book was published under two titles, one being "A Project for a Scientific Psychology."

53. Claude Lévi-Strauss, *Tristes Tropiques: An Anthropological Study of Primitive Societies in Brazil* (New York: Atheneum, 1961), 61.

54. See David Hume, *A Treatise of Human Nature* (Oxford: Oxford University Press, 1888), xvii. As Hume announces at the onset of the *Treatise*, "'Tis easy for one of judgment and learning to perceive the weak foundation even of those systems that have obtained the greatest credit and have carried their pretensions highest to accurate and profound reasoning. Principles taken upon trust, consequences lamely deduced from them, want of coherence in the parts, and of evidence in the whole, these are everywhere to be met with in the systems of the most eminent philosophers and seem to have drawn disgrace upon philosophy itself."

55. See also Stephen J. Gould, *Time's Arrow, Time's Cycle: Myth and Metaphor in the Discovery of Geological Time* (Cambridge, Mass.: Harvard, 1987).

56. Similar struggles emerge when sociologists attempt to say what Max Weber meant in his famous study *The Protestant Ethic and the Spirit of Capitalism*, trans. Talcott Parsons (New York: Scribner, 1958). Many have criticized Weber's claim that Protestant religious ideas drove some Germans and English into capitalist habits that helped them trump Catholics in France and Italy. Saying exactly what Weber means by the "spirit of capitalism," and showing that he meant that religious ideas affected the rise (or absence) of capitalism, is not easy. For example, Weber says that "the spirit of capitalism (in the sense we have attached to it) was present *before* the capitalistic order," 55 (emphasis mine). See also George Becker, "The Continuing Path of Distortion: The Protestant Ethic and Max Weber's School Enrollment Statistics," Unpublished manuscript.

57. Max Weber, *The Protestant Ethic and the Spirit of Capitalism and Other Writings by Max Weber*, ed. and trans. Peter Baehr and Gordon C. Wells (New York: Penguin Books 2002), xxxiii, xxiv.

58. W. D. Falk, "Hume on Is and Ought," *Canadian Journal of Philosophy* 6 (1976): 359–78; and Paul Guyer, *Kant and the Claims of Taste* (Cambridge, Mass.: Harvard University Press, 1979). See also Ted Cohen and Paul Guyer, eds., *Essays in Kant's Aesthetics* (Chicago: University of Chicago Press, 1982).

59. *Stanford Encyclopedia of Philosophy*, "David Hume," at http://plato.stanford .edu/entries/hume/ (accessed November 14, 2008). "Intellectualism and sentimentalism seem to be exhaustive alternatives, ways of characterizing the ancient debate as to whether reason or passion is, or should be, the dominant force in human life. Hume saw that both approaches capture important aspects of human nature but that neither tells the whole story. We are active and reasonable creatures. A view that mixes both styles of philosophy will be best, so long as it gets the mixture right."

60. Josef Breuer and Sigmund Freud, *Studies in Hysteria*, vol. 2 of 24, *The Standard Edition of the Complete Psychological Works of Sigmund Freud*, ed. J. Strachey (London: Hogarth Press, 1895), 46. *Hysteria* refers neither to a discreet disease entity nor to a well-defined set of behaviors. Even clinicians who share a particular tradition differ widely in their interpretations of patient behaviors. Breuer diagnosed Anna O. as a hysteric with psychotic episodes. Others have diagnosed her as outright psychotic (S. Reichert, "A Re-examination of 'Studies on Hysteria,'" *Psychoanalytic Quarterly* 25 [1956]: 162); mixed neurotic character, severely hysteric (R. Karpe "The Rescue Complex in Anna O's Final Identity," *Psychoanalytic Quarterly* 30 [1961]: 23); transient melancholic due to her father's death (G. H. Pollock, "The Possible Significance of Childhood Object Loss in the Josef Breuer–Bertha Pappenheim (Anna O.)–Sigmund Freud Relationship (I. Josef Breuer)," *Journal of American Psychoanalytic Association* 16, no. 4 [1968]: 711–39.); and as a unique case in no sense a hysteric (Henri Ellenberger, *The Discovery of the Unconscious: The History and Evolution of Dynamic Psychiatry* [New York: Basic Books, 1970]). The fact that Anna was not cured by Breuer's chimney sweeping, and that she suffered hallucinatory states for many years after, challenges all diagnoses (E. Jones, *The Formative Years and the Great Discoveries, 1856–1900*, vol. 1, *The Life and Work of Sigmund Freud* [New York: Basic Books, 1953], 224).

61. Breuer and Freud, *Studies*, 36.

62. Breuer and Freud, *Studies*, 36.

63. Breuer and Freud, *Studies*, 295.

64. Breuer and Freud, *Studies*, 286.

65. Breuer and Freud, *Studies*, 268 (emphasis original).

66. Breuer and Freud, *Studies*, 268.

67. S. Bernfeld, "Freud's Earliest Theories and the School of Helmholtz," *Psychoanalytic Quarterly* 13 (1944): 341–62.

68. S. Freud, *The Origins of Psycho-Analysis* (London and New York: Imago, 1954). Partly including "A Project for a Scientific Psychology," in *Standard Edition*, 1, 175.

69. Breuer and Freud, *Studies*, 272 (emphasis original).

70. Breuer and Freud, *Studies*, 265.

71. Breuer and Freud, *Studies*, 282.

72. Breuer and Freud, *Studies*, 283, 302–303.

73. Breuer and Freud, *Studies*, 305.

74. Bruce E. Wampold, *The Great Psychotherapy Debate: Models, Methods and Findings* (Mawah, N.J.: Erlbaum, 2001).

75. Wampold, *Great Psychotherapy Debate*, 204.

76. Wampold, *Great Psychotherapy Debate*, 204.

77. Jerome Frank, *Persuasion and Healing: A Comparative Study of Psychotherapy* (Baltimore: Johns Hopkins Press, 1991).

78. Wampold, *Great Psychotherapy Debate*, 206.

79. See Lévi-Strauss, *Tristes Tropiques; La Pensee Sauvage* (Paris: Plon, 1962). See also his *Totemism*, trans. R. Needham (Boston: Beacon, 1962). And see Volney Gay, "Ritual and Psychotherapy: Similarities and Differences," in *Religious and Social Ritual*, ed. V. De Marinis and M. Aune (New York: State University of New York Press, 1996), 217–34.

80. Thus, on whether or not Baruch Spinoza (1632–1677) identified the mind with the brain, see Harold F. Hallett, "On a Reputed Equivoque in the Philosophy of Spinoza," *Review of Metaphysics* 3 (1949), and online at http://caute.net.ru/spinoza/aln/equivoq.htm (accessed November 14, 2008). Hallett says, "Among the various ambiguities that have been invented (and I use the word equivocally) in the philosophy of Spinoza and have been held to be fatal in respect of this or that part of it, or of the whole, none seems to have been so generally, and so undoubtingly, noted as his identification of the bodily correlate of the human mind with its physical object in Ethics II, xiii. And certainly no error (if it is an error) could be more fundamentally fatal to the whole speculation; an equivocal use of the term *idea* as at once the mental correlate of some neural or physiological state of the body of the percipient, and also the *essentia objectiva* of a thing extrinsic to that body—involving the simple identification of the human body with the object of the human mind that animates it—would seem to be a confusion at the source, infecting the whole system."

81. See Randall Lehmann Sorenson, "Psychoanalytic Institutes as Religious Denominations: Fundamentalism, Progeny, and Ongoing Reformation," *Psychoanalytic Dialogues* 10, no. 6 (2000): 847–74. In unfortunate cases, training becomes a form of denominationalism.

82. See H. W. Loewald, "Psychoanalysis as an Art and the Fantasy Character of the Psychoanalytic Situation," *Journal of the American Psychoanalytic Association* 23, no. 2 (1975): 277–99.

4

Neuroscience, Theory of Mind, and the Status of Human-Level Truth

Edward Slingerland

MANY COMMITTED SCIENTISTS who nonetheless continue to see the value and beauty of traditional religious belief have argued that science and religion are simply talking about different things. In the famous characterization of the late Stephen Jay Gould, they deal with "nonoverlapping magisteria."[1] I argue that, in fact, the scientific model of the self and the traditional religious models of the self *are* in fundamental tension and that much of the continuing resistance to integrating findings from neuroscience and other areas of the sciences into the study of religion stems from this fact. The neurobiological view of the self—grounded in a physicalist model of the body-mind as a product of evolution—fundamentally conflicts with our intuitive "folk" notions of who and what we are. The late Francis Crick spent the latter part of his career exploring the neuroscience of consciousness. He labeled *the astonishing hypothesis* the idea that "*all* aspects of the brain's behavior are due to the activities of neurons"[2]—that is, that consciousness can ultimately be reduced to a physical chain of firing neurons. It is, in fact, more than astonishing: the physicalist view of the human self and the human mind is alien and profoundly disturbing. The implications of this picture of the human mind/self/soul are summed up vividly and succinctly in a quotation from the Italian neuroscientist Giulio Giorelli: "Yes, we have a soul, but it is made up of many tiny robots."[3] Whatever view one ultimately adopts vis-à-vis this thoroughly materialist view of the self (and I suggest a slightly alternate view in the following), these ideas have to be grappled with, because, strange as they are, it is difficult to see what choice we have once we take the decisive step of giving up our belief in a Cartesian ghost in the machine—of believing, to put a finer point on it, in magic. Given the apparent explanatory superiority of physical-ism, and the absence of empirically viable alternatives to it, it is hard to avoid the conclusion we are "little robots" all the way down.

At first blush this would seem to imply that humanists are now out of a job. The apparent dependence of the humanities on a dualistic model of the self is made quite clear in German, where they are referred to as the *Geisteswis-senschaften*—the sciences of the *Geist* (mind, spirit, ghost). What need is there for the *Geisteswissenschaften* if there is no such thing as *Geist?* Below I discuss the issues of reductionism, levels of explanation, and innate human cognitive resistance to physicalism and finally draw from the work of Charles Taylor to sketch out a somewhat more finessed conception of realism than one tends to encounter among scholars with strong natural-scientific bents. I hope to convince you that we do not need to retreat to some vaguely conceived form of dualism in order to do justice to both our sense of what is empirically plau-sible and our deep intuitions about the world—that we can acknowledge the neuroscientific model of the self while still living and working in an environ-ment rich with human meaning.

The Bogeyman of Reductionism

One of the standard critiques of the physicalist view of the human being is that it is overly reductionistic, an adjective that—like the equally damning charge of realist one hears quite frequently in humanistic circles—functions nowadays primarily as a vague term of abuse. To begin with, it is important to realize that any truly interesting explanation of a given phenomenon is in-teresting precisely *because* it involves reduction of some sort—tracing causa-tion from higher to lower levels or uncovering hidden correlations. As Steven Pinker puts it, the difference between reductive and nonreductive explanation is "the difference between stamp collecting and detective work, between sling-ing around jargon and offering insight, between saying something just is and explaining why it had to be that way as opposed to some other way it could have been."[4] We are not satisfied with explanations unless they answer the why question by means of reduction: by linking the *explanandum* to some deeper, hidden, more basic *explanans.*

This is why the way that even traditional humanist scholars go about their work is already essentially reductionistic. My first monograph was a study of five early Chinese thinkers.[5] I argued that these thinkers shared a common spiritual ideal, that there was a tension internal to this spiritual ideal that motivated much of their theorizing about human nature and self-cultivation, and that looking at these thinkers through the lens of this tension explained the development of early Chinese thought—and indeed the later trajectory of East Asian religious thought—in a uniquely satisfying and revealing way. I did not simply write my own version of an early Chinese philosophical text.[6] Rather, I *reduced*, trying to show how five apparently disparate texts could, in fact, be seen as motivated by a single, deeper, shared goal and common conceptual tension. Now, some of my colleagues think that my book is a *bad* reduction, that it glosses over important distinctions or distorts certain positions in order to make them fit this new narrative. But no one criticizes the project as reductionistic *tout court*. Reduction is what we do as scholars, humanistic or otherwise, and when someone fails to reduce, we rightly dismiss their work as trivial, superficial, or uninformative.

When the deeper principles behind things are poorly understood—that is, when lower levels of causation underlying phenomena we are interested in explaining are not accessible to our prying—we are often forced to invent vague, place-holder entities to stand in for the missing information. Sometimes we are aware that this is what we are doing. For instance, Mendel could reason about the inheritance of traits without knowing how information about them was physically instantiated or transmitted, and Darwin could similarly map out the implications of natural selection without any clear conception of the substrate of inheritance. In such cases there is an implicit faith that the lower-level entities and processes will eventually be specified; if not, the theory may have to be abandoned. A discipline can find itself in a dead-end, however, when it has postulated vague, place-holder entities without realizing that this is what it is doing—when it takes these unspecified and unknowable entities or faculties to have genuine explanatory force. Gottfried Leibniz felt that this was the case with regard to the field of human psychology at his time, which relied heavily on such concepts as the intellect or understanding. He notes that to invoke concepts like these is to merely save appearances "by fabricating faculties or occult qualities . . . and fancying them to be like little demons or imps that can, without ado, perform whatever is wanted, as though pocket watches told the time by a certain horological faculty without needing wheels, or as though mills crushed grain by a fractive faculty without needing anything in the way of millstones."[7]

Nietzsche similarly mocked Kant for thinking that he was saying something substantive about the human capacity to create and feel the force of moral imperatives with his analysis of synthetic a priori judgments:

"How are synthetic judgments a priori *possible*?" Kant asked himself. And what really is his answer? "By virtue of a faculty" [*Vermöge eines Vermögens*][8]—but unfortunately not in five words, but so circumstantially, venerably, and with such a display of German profundity and curlicues that people simply failed to note the comical *niaiserie allemande* involved in such an answer. People were actually beside themselves with delight over this new faculty. . . . "By virtue of a faculty," [Kant] had said, or at least meant. But is that an answer? An explanation? Or is it not rather merely a repetition of the question? How does opium induce sleep? "By virtue of a faculty," namely the *virtus dormitiva*, replies the doctor in Molière,

> Because it contains a sleepy faculty
> Whose nature is to put the senses to sleep.[9]

"Such replies belong in comedy," Nietzsche concludes, and so should we.

The force of the argument of cognitive scientists and evolutionary psychologists who are pushing for vertical integration between the humanities and the natural sciences—for integrating the neuroscientific model of the self into our picture of the human—is that the humanities have yet to genuinely free themselves from this sort of Tartuffery[10] and continue to rely on impressive-sounding but explanatorily empty entities and faculties. For instance, John Tooby and Leda Cosmides note that most humanistic and social-scientific models are based on a picture of human nature as a blank slate that gets filled up by means of learning, which is about as helpful an explanation as that opium produces sleep by means of its sleepy faculty: Learning—like culture, rationality, and intelligence—is not an explanation for anything but is rather a phenomenon that itself requires explanation. In fact, the concept of learning has, for the social sciences, served the same function that the concept of protoplasm did for so long in biology.[11]

Just as this mysterious protoplasm turned out to consist of a collection of distinct intricate structures with specific functions, Tooby and Cosmides argue, so will words like *learning, intelligence,* and *rationality* turn out to be blanket terms for what are really a variety of specific, modular, evolved cognitive processes that allow human beings to selectively extract and process adaptively relevant information from the world.

Of course, it is obvious that the larger meaning of a particular wink—Why is *this person* winking at me? What should I do?—is embedded in a set of long, complex stories and that for the unpacking and analysis of these stories, we require the higher-level expertise of anthropologists, novelists, and historians. Such humanistic work, however, should not be seen as occurring in an explanatory cloud-cuckoo land, magically hovering above the mundane world of physical causation. Despite their variety and disunity, the various disciplines of the natural sciences have managed to arrange themselves in a rough explanatory hierarchy, with the lower levels of explanation (such as physics) setting limits on the sorts of explanations that can be entertained at the higher levels (such as biology). In order to move forward as a field of human inquiry, the humanities need to plug themselves into their proper place at the top of this explanatory hierarchy, because the lower levels have finally advanced to a point where they have something interesting to say to the higher levels. Human-level meaning emerges organically out of the workings of the physical world, and we are being reductive in a good way when we seek to understand how these lower-level processes allow the higher-level processes to take place.

From Physicalism to the Humanities: Levels of Explanation

Having sent the bogeyman of reductionism back to its cave, it is now possible to talk about good and bad forms of reductionism—because, of course, it is really greedy or eliminative reductionism that most humanists have in mind when they bandy about this charge. In order to distinguish productive, explanatory reductionism from crudely eliminative reductionism, it is important to get some clarity about the heuristic and ontological status of entities at various levels of explanation.

Levels of Explanation and Emergent Qualities

Although no evolutionary psychologist or neuroscientist would purport to be an eliminative reductionist—and all give lip service to the idea that higher levels of explanation can feature emergent qualities not present at the lower levels—there is a common tendency to nonetheless privilege the material level of explanation: we are really just mindless robots or physical systems, no matter how things might appear to us phenomenologically. There are some very good reasons for this privileging of lower levels of explanation. To begin with, the physicalist stance has proven extremely productive, allowing such dramatic technological developments as supercomputers and pharmacological treatments for mental illnesses. Moreover, there is an a priori reason for giving precedence to the physical: the structure of the various upper levels of explanation emerges out of and depends on the lower levels, so the lower levels *are* causally privileged in this way. Molecules form and behave in accordance with more basic principles that govern both inorganic and organic substances, which means that a hypothesis in molecular biology that violates well-established physical chemistry principles is either wrong or a reason for us to rethink our physical chemistry.

It is equally the case, however, that as we move up the explanatory chain we witness the emergence of what appear to be new entities, which possess their own novel and unpredictable organizational principles. One of the clearest ways to get a grasp on how emergent properties emerge, as well as what their ontological status might be, is to turn to a model discussed by Daniel Dennett, the Game of Life developed by the mathematician John Conway.[12] The game is played on a two-dimensional grid—originally a checkerboard—with each cell of the grid arbitrarily set to an initial "alive" (COUNTER) or "dead"

(NO COUNTER) state. In the first round, the initial grid is then transformed in accordance with a simple set of rules:

> Survivals. Every counter with two or three neighboring counters survives for the next generation.
>
> Deaths. Each counter with four or more neighbors dies (is removed) from overpopulation. Every counter with one neighbor or none dies from isolation.
>
> Births. Each empty cell adjacent to exactly three neighbors—no more, no fewer—is a birth cell. A counter is placed on it at the next move.[13]

There are two important things to note about the Game of Life. The first is that it is an obviously deterministic world: once the initial state is set, the manner in which this state will transform is determined algorithmically and is perfectly predictable. As Dennett puts it, "when we *adopt the physical stance* toward a configuration in the Life world, our powers of prediction are perfect: there is no noise, no uncertainty, no probability less than one."[14] The second feature of the game is how quickly it gives rise to emergent phenomena: as counters are removed or placed on the board with each iteration of the algorithm, specific and coherent shapes begin to emerge, and these shapes often appear to begin "behaving" and "interacting" in novel ways. Although many initial configurations either "die out" quickly, or lead to stable configurations that cease to change (what Conway called *still lifes*), others are more dynamic. Here are three of the sample initial states and their outcomes described by Gardner in his original article (figure 4.1):[15]

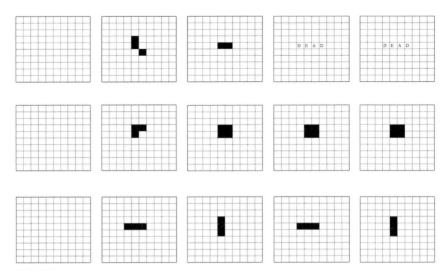

FIGURE 4.1
The Fate of Three Initial Configurations in the "Game of Life"

The first Life World quickly dies out, the second freezes into a static block, and the third results in a "blinker," which appears to perpetually flip back and forth between a horizontal and vertical orientation.

After being introduced by Gardner's articles in *Scientific American*, the Game of Life was quickly adapted to the computer, with digital grids blinking ON and OFF replacing the chess squares and counters. The massively greater computational power and speed of the digital computer allowed huge grids and an infinite number of randomly generated initial states to be run, resulting in Life Worlds of greater and greater complexity, as well as a growing menagerie of Life World entities. One of the most interesting and yet simplest is the glider, an apparently self-propelled configuration that moves diagonally across the Life World grid, often colliding with and interacting with other emergent entities. For an example, see the five-generation snapshot of two gliders approaching each other from opposite corners of the Life World and colliding, with only one emerging alive (figure 4.2).[16]

As William Poundstone observes concerning basic glider behavior in the simulations that he ran, "A glider can eat a glider in four generations. Whatever is being consumed, the basic process is the same. A bridge forms between the eater and its prey. In the next generation, the bridge region dies from overpopulation, taking a bite out of both eater and prey. The eater then repairs itself. The prey usually cannot. If the remainder of the prey dies out, as with the glider, the prey is consumed."[17]

It is revealing how easily and quickly observers of the Life World begin to attribute both bounded identities and intentionality to what is, we have to remind ourselves, merely a pattern of squares turning on and off in accordance with a basic algorithm. It is also important to note how, once we become interested in emergent patterns such as gliders, we begin to formulate new generalizations—completely unrelated to the simple, rigid algorithm that is the basis of the entire world—to describe and predict their "behavior." As Dennett observes, referring to this new level of description as the *design level*:

> Notice that something curious happens to our "ontology"—our catalogue of what exists—as we move between levels. At the physical level there is no motion, just ON and OFF, and the only individual things that exist, cells, are defined by their fixed spatial locations. At the design level we suddenly have the motion of persisting objects. . . . Notice, too, that whereas at the physical level there are absolutely no exceptions to the general law, at this level our generalizations have to be hedged: they require "usually" or "provided nothing encroaches" clauses.[18]

The Life World thus provides us with a wonderfully simple model of how a completely deterministic, clearly algorithmic system can give rise to emergent, higher levels of reality—the design level of gliders and blinkers—that function in a manner not obviously predictable from basic, lower-level laws. There is

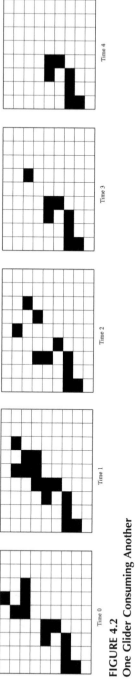

FIGURE 4.2
One Glider Consuming Another

no way, given an initial state and the guiding algorithm, that one could predict that entities called gliders will arise and will tend to "behave" in certain ways: you need to actually *run* the Life World simulation on a computer and see what happens. With the advent of the Internet, increasingly complex Life World computer simulations have become extremely popular, and entire menageries of wild entities—glider guns that shoot out steady streams of new gliders, logic gates, prime-number generators—have been discovered, none of which could have been predicted by Conway when he originally conceived of his simple checkerboard solitaire game. And yet we can readily acknowledge that none of these entities really exist: because we have been shown how the system is constructed, we are capable of pulling back from the drama of glider predator or the operations of logic gates to remind ourselves that all of this is merely the blinking ON and OFF of fixed, square cells.

To relate the Life World model and the lessons it teaches us to more obviously relevant concerns, we see a similar shift in ontology and variation in degrees of certainty as we move up through the levels of the reality studied by the natural sciences. The field of organic chemistry is based on principles that emerge at the level of organic molecules, which cannot be fully predicted from the perspective of physical chemistry. Generalizations about the behavior of organic molecules also tend to be less precise, and more subject to *ceteris paribus* clauses, than the principles of physical chemistry. Similarly, no amount of intimacy with quantum-mechanical principles will allow one to even begin to predict the behavior of macrolevel solid objects. As Hilary Putnam famously observed, there are entire fields of human knowledge, such as geometry, that emerge only once we reach the level of macro objects.[19] The fact that a square peg fifteen-sixteenths of an inch per side will fit through a one-by-one-inch square hole but not circular hole one inch in diameter, is a function of the peg's geometric properties; referring to the properties of the molecules that make up the pegs or the materials from which the holes are drilled would be heuristically useless.

Even within what are sometimes assumed to be single fields, such as biology, there exist multiple levels of explanation, with structures at the higher levels being in no clear way predictable from or simply reducible to the levels below. As E. O. Wilson notes, the organization of social-insect societies, such as termites and ants, is made possible only because of an unusual reproductive strategy, haplodiploidy, whereby fertilized eggs produce females and unfertilized eggs produce males. How or why this strategy may have evolved is open to speculation, but whatever its cause it led to the development of a very unique form of social organization. This is because, with haplodiploidy reproduction, sisters are more closely related to each other than mothers are to their daughters, which makes the strategy of females becoming a sterile slave caste dedicated to the rearing of sisters a potentially advantageous one. As Wilson observes, "The societies of wasps, bees, and ants have proved so successful that they dominate and alter most of the land habitats of the Earth. In the forests of Brazil, their assembled forces constitute more than 20 percent of the weight of all land animals, including nematode worms, toucans, and jaguars. Who could have guessed all this from a knowledge of haplodiploidy?"[20]

When it comes to closely associated levels, causal predictive power can also move in either direction of the vertically integrated chain. Lindley Darden and Nancy Maull, for instance, provide an illuminating account of the historical development in two closely related fields, cytology (cell biology) and genetics, noting that throughout this history "predictions went in both ways," with discoveries at each level focusing attention in the other level in ways impossible to anticipate in advance.[21]

This mutual dependence and interaction of levels of explanation is taken for granted in the natural sciences and is in fact one of the guiding principles driving natural scientific inquiry. The challenge involved in integrating neuroscience and the study of religion is hooking the various levels of explanation employed in the humanities into their proper place at the top of this causal explanatory chain. To oversimplify a bit by bringing this back to our Life World example, humanists are currently in the position of trying to explain emergent-entity behavior as a mysterious, sui generis phenomenon and react with hostility or disdain to the suggestion that, for instance, blinking squares have anything to do with the free and autonomous workings of the glider spirit. There *is*, as we have seen, something novel and heuristically useful about the principles of glider-level behavior. Moreover, real-life natural scientists, who have as yet only a partial understanding of a physical world inconceivably more complex than Conway's Life World, would no doubt find accounts of glider-level phenomena very helpful in their struggle to decipher these real-world, lower-level mechanisms. These two sets of researchers need to begin comparing notes, though, with the recognition that both have something important to bring to the table. Long before the emergence of sociobiology and evolutionary psychology, John Dewey argued for a "principle of continuity" between the various fields of human inquiry and the natural sciences, noting that there was no reason to see a "breach in continuity between operations of inquiring and biological operations and physical operations."[22] He also saw that continuity did not mean ignoring the existence of different levels of explanation, each with their own guiding principles. Continuity between such human-level phenomena as rationality and more basic-level phenomena as human biology means that "rational operations *grow out of* organic activities, without being identical with that from which they emerge."[23] Let us now turn to a discussion of how some human-level phenomena that seem extremely real and important to humanists could plausibly emerge from lower-level phenomena, as well as how precisely we should understand the ontological status of these various levels.

The Emergence of Free Will and Intentionality

How could free will emerge from a deterministic universe without our having to postulate an entirely new type of entity? In the past decade or so a veritable cottage industry has sprung up dedicated to physicalist, neurobiological accounts of consciousness, free will, and human intentionality, and providing even a brief survey of this vast literature would require a book in itself.[24] In the interest of brevity, I will focus on one of the more prominent accounts, that of Daniel Dennett, fleshing out his view with some observations from other researchers in the field.

Dennett begins by observing that unusually complex survival machines are, by their very nature, prone to gradually acquire their own goals and motivations. He asks his reason to consider possible strategies available to someone keen on preserving him or herself in a cryogenic chamber into the twenty-fifth century, faced by the need to plan for energy needs, safety from natural disaster, et cetera, with really only a vague sense of what challenges this chamber will face over the next few centuries.[25] A fixed chamber well-supplied with what it could foreseeably need (the *plant* strategy) is one possibility, but a major drawback is that it cannot move if resources run out or destructive changes occur to its habitat. A mobile chamber option (the *animal* strategy) is more flexible: it can be designed to actively seek out resources to keep itself running and also to perceive and avoid potential dangers. One catch is that you are unlikely to be the only person to have thought of this strategy, so a major danger facing your cryogenic chamber is no doubt going to be other cryogenic chambers, all competing for the same resources and possibly cannibalizing each other for parts and fuel. Basically, the best strategy in the face of these design constraints is to program your cryogenic machine with a set of basic *desiderata*, basic capacities for acquiring and processing information from the world that will be relevant to its mission, and then let it go. This robot will by necessity be capable of exhibiting self-control of a high order. Since you must cede fine-grained real-time control to it once you put yourself to sleep, you will be as remote as engineers designing an autonomous space probe. As an autonomous agent, it will be capable of deriving *its own* subsidiary goals from its assessment of its current state and the import of that state for its ultimate goal (which is to preserve you till 2401). These secondary goals, which will respond to circumstances you cannot predict in detail (if you could, you could hard-wire the best responses to them), may take the robot far afield on century-long projects, some of which may be ill advised, in spite of your best efforts. Your robot may embark on actions antithetical to your purposes, even suicidal, having been convinced by another robot, perhaps, to subordinate its own life mission to some other. All the preference it will ever have will be offspring of the preferences you initially endowed it with, in hopes that they will carry you into the twenty-fifth century, but that is no guarantee that actions taken in light of the robot's descendent preferences will continue to be responsive, directly, to your best interests.[26]

The robot, of course, is a Dawkins-esque survival machine, and the planner of this robot is the genes. This thought-experiment makes two important points. One is that, although the basic motivational and behavioral profile of a survival machine should be roughly comprehensible in terms of its original programmed operating instructions, there is no guarantee that its actual motivations or behavior will not deviate from these instructions in unpredictable ways.

The ability to draw connections between schemas from different domains—a hallmark of what Steven Mithen has referred to as "cognitive fluidity"[27]—is a powerful feature of the variety of survival machine known as homo sapiens, and the genes that included it their design have done fairly well. They could never have predicted, however, that the quite-reasonable factory-installed desire for cleanliness or physical purity—an excellent mechanism for assuring that your survival machines avoid taking in fuel that will harm their operating systems—could be drawn upon and projected onto the realm of sexual relations. In the course of human history, this mapping has independently and repeatedly resulted in certain survival machines deciding that *sex*, their very raison d'être, is unclean and should be avoided by those machines desiring to remain "spiritually" clean. Although this is obviously a bad outcome for the particular genes residing in those survival machines, the payoff of cognitive fluidity must be high enough that their terrible fate has been historically outweighed by the success of copies of them residing in other individual survival machines fortunate enough not to have embraced this particular metaphor.

The second important point of the cryogenic robot scenario is that it is not at all surprising that robots built by genes to get them into the next generation could possess extremely broad degrees of behavioral freedom and unpredictability as basic design features. In other words, "freedom evolves"[28]: behavioral self-control and the ability to reassess motivations and goals emerge gradually as the complexity of survival machines increases, in a manner that we can perceive quite clearly from surveying the varying degrees of self-awareness and intelligence among the profusion of species that have made it this far in our world. Our intuitive sense that we possess free will, deliberate over choices, are torn in different directions, and ultimately make decisions that could have gone either way is not at all mistaken. We just need to recognize that this free will is not some magical, inexplicable quality existing completely outside the chain of causation, and possessed only by human souls. It is, rather, an emergent quality that is helpful for understanding the behavior of *any* survival machine functioning at a sufficient level of complexity.

Weak versus Strong Emergence: Blocking the Move to Mysterianism

We are familiar with how the process of evolution and natural selection has produced more and more complex feeding and fleeing machines, working at many different layers in the food chain, as well as wildly diverse strategies of hunting, mating, parenting, and social organization.[29] Very crude survival machines are built to sense temperature and inorganic nutrient gradients and adapt their movements and simple feeding behaviors accordingly. More complex ones are then built to take advantage of the work already done by these simple machines in concentrating diffused inorganic nutrients in one valuable package: they are the first predators and require more complicated sensory and behavioral programming to track down and capture their prey. The prey, in turn, become more complex in response to this pressure, acquiring the ability to detect and evade predators. At a certain point in this process of exponentially increasing complexity, it became more efficient for survival machines to have built into them the set of heuristics Dennett refers to as the "intentional stance"[30] to predict and respond to the behavior of other complex survival machines—trying to rely on the physical stance, still helpful for dealing with simple rocks and trees and coconuts, simply wasn't fast enough. This set of fundamental intuitions has come to be referred to by cognitive scientists as theory of mind (ToM),[31] which governs our interactions with other people—causing us to paint mental properties onto what, when we take a step back, we can acknowledge is really just a sequence of physical states: pupils dilating, limbs moving, jaw and lip muscles contracting in a certain sequence. We cannot help but see the physical movements of other human beings as being caused by the workings of a nonmaterial mind, full of desires, thoughts, beliefs, and fears. Once theory of mind developed, complex survival machines such as ourselves could not help but begin viewing other survival machines as fearing, wanting, and liking. Especially in the mental repertoire of the most complex of these survival machines—whose adaptive environment consists primarily of other conspecifics rather the dumb world of things—the concepts of behavioral "choice" and "freedom" became very useful heuristics.

This is the story of the emergence of human free will presented by weak emergentists: freedom as an emergent quality of very complex machines, with there being no point in this chain of increasing complexity where something magic happens. Like the category of glider, we can have new kinds emerging at each new level of explanation—kinds that are heuristically indispensable, that present themselves irresistibly to the human mind as crucial features of causal explanation, and yet that are not composed of novel stuff. As Michael Arbib puts it, referring to folk psychology concepts or person-talk, they "are useful for encapsulating meaningful patterns of what our brains can do, but not as describing a distinct reality."[32]

Opposing the weak-emergence stance are the advocates of various forms of strong or ontological emergence. These include, of course, old-fashioned substance dualists, who claim that mind and matter are two independent ontological realms. Descartes is the classic exponent of this position, and—despite the ill repute into which Cartesianism has fallen in recent decades—full-blown substance dualists are still fairly thick on the ground. Those who adhere to traditional religious models of the self are obviously and explicitly dualistic in this sense, but even many otherwise committed secular physicalists appear to continue to hear the siren call of substance dualism.[33]

A more updated, but—to my mind, at least—indistinguishable position is so-called property dualism, which argues that things like human qualia are ineffable and possess strongly emergent properties. *Qualia* is a technical-sounding philosophical term for what is, in fact, a quite folksy idea: there is a "what-it-is-like-ness" to my conscious experience that is immediately and exclusively accessible only to myself, and this special qualitativeness is what would be left out of any third-person description of my experience. Thomas Nagel provided the classic statement of this position in his famous 1974 essay "What Is It Like to Be a Bat?" where he argued that, essentially, we can never answer that question: we are not bats, and no matter how much third-person descriptive knowledge we accumulate about bat behavior and physiology, we can never have access to the first-person (first-mammal?) qualia of bat consciousness.

The qualia argument, as intuitively appealing as it is, has always seemed to me more an item of faith or bald assertion than an argument, per se, and both Daniel Dennett and Hilary Putnam—to take just two examples—have formulated eloquent and convincing critiques of the concept.[34] To begin with, despite its initial intuitiveness, it is hard to know exactly what it is that a qualia might be. Dennett notes that the distinctive richness of qualia is linked in most people's minds to our experiential *je ne sais quoi*—the fact that the lived details of our experience seem to defy adequate verbal description. In "Quining Qualia" (1988), he asks us to consider a Jell-O box that has been torn in half, with one half forming a complexly contoured shape we will label *M*. Providing a full and adequate description of *M* is virtually impossible, and for practical purposes the only way to identify *M* is to rely on the unique *M*-detector—that is, the other half of the box. The shape of *M*, Dennett notes, "may *defy description*, but it is not literally ineffable or unanalyzable; it is just extremely rich in information. It is a mistake to inflate practical indescribability into something metaphysically more portentous."[35]

Hilary Putnam makes a related point in his discussion of the so-called zombie problem in philosophy of consciousness, which in its contemporary form can be traced back to Descartes' apparently quite-unhealthy obsession with automatons[36] and William James's musings about an "automatic sweetheart" indistinguishable from a flesh-and-blood woman.[37] What are we to make of a machine/creature that looks exactly like a normal human being, talks exactly like a normal human being—in short, is indistinguishable from a normal human being on the surface—but, in fact, completely lacks consciousness and, therefore, the experience of qualia? Putnam argues quite forcefully that the zombie scenario is in fact a nonproblem, caused by unexamined and ultimately untenable dualist assumptions. "In the absence of soul talk," he notes, "the very idea that our mental properties might be 'subtracted' from us without disrupting our bodies or altering our environments" is basically "unintelligible."[38] It is unintelligible, that is, in light of the physicalist assumptions that guide our interactions with the rest of nonhuman reality, which—in light of their success to date—we have no principled reason to avoid extending to human consciousness. Qualia are supposedly something above and beyond the functioning of one's embodied mind as it goes about its work, but we have no reason to believe that what we experience as consciousness *is* anything other than the embodied mind going about its work.

Second, the privilege given to the first-person perspective is a natural human prejudice, but a built-in bias—no matter how powerful—is not, in and of itself, adequate evidence for the existence of an ontologically distinct realm. At a crucial point in his 1974 article, Nagel asks rhetorically, "Does it make sense . . . to ask what my experiences are *really* like as opposed to how they appear to me?"[39] This is really the key to his argument. Our bad-argument radar should go off when a philosopher's core position is expressed as a rhetorical question, and we can defang the rhetorical bite of Nagel's query by answering it in the affirmative. Of course it makes sense to ask what my experiences are really like as opposed to how they seem to me: the idea that third-person accounts can never get to the essence of first-person phenomenology is nothing more than a feeling that we have—an expression of the illusion of self-unity and complete self-transparency that is created by our fragmented and multilevel mind. Work coming out of cognitive and social psychology is calling into question our folk intuition that we possess a unified locus of consciousness: a fully self-aware homunculus supposedly pulling the levers and running the show.[40] To answer Nagel's supposedly unanswerable question, then, we can say that a measure of the difference between what one's experiences are really like and how they appear introspectively can, as Jonathan Schooler and Charles Schreiber put it, be derived from "the degree to which self-reports systematically covary with the environmental, behavioral, and physiological concomitants of experience."[41] We can acknowledge the important role that qualia play in the experienced economy of the human psyche without attributing to them mysterious powers or special ontological status.

Perhaps the most prominent and prolific critic of a thorough-going materialist view of human consciousness is John Searle, who argues that no third-person, purely physicalist account can capture the "original intentionality" or "ontological subjectivity" that is an essential characteristic of human consciousness.[42] Searle's argument at times seems structurally identical to the qualia argument, although it is somewhat difficult to get a real handle on his position. To begin with, in his more recent work he is an outspoken critic of substance dualism and more explicitly mysterian positions such as the one espoused by Nagel. He now characterizes himself as a biological naturalist[43] and purports to be a physicalist in the sense that he does not believe there is any substance in the universe over and above physical things: conscious states have "absolutely no life of their own, independent of the neurobiology."[44] His position sometimes sounds like what we would characterize as weak emergence: "roughly speaking, consciousness is to neurons as the solidity of pistons is to the metal molecules."[45] On the other hand, he still wants to maintain that conscious states are in some way "ontologically irreducible" to physical states: the emergent property of piston solidity is both causally and ontologically reducible to the characteristics of metal molecules, whereas consciousness is causally, but not ontologically, reducible to neuron activity. Why is this? Because we "just know" that conscious states exist, in that we experience them all the time. Searle is relying here on Descartes' basic *cogito* argument: "If it consciously seems to be that I am conscious, then I am conscious. I can make all sorts of mistakes about the contents of my conscious states, but not in that way about their very existence."[46] He thus concludes that there is something special about consciousness that warrants our postulating the existence of a unique "first-person ontology" alongside the ordinary "third-person ontology" that is adequate for characterizing the rest of the physical world.

Unless Searle is using the word *ontology* in a radically idiosyncratic sense, however, he seems to be falling back here into precisely the sort of substance-dualist position that he elsewhere argues has outlived its usefulness. To get a sense of how original intentionality or ontological subjectivity requires a strong sense of emergence, let us return to the Life World example. As we get increasing levels of complexity, new heuristics for predicting what patterns of squares are going to appear in the next generations become available to us: blinkers will switch from a horizontal to vertical orientation, gliders will try to eat other gliders when they collide, et cetera. Bringing this back home to the biological world, life began when some molecules managed to begin replicating themselves. Soon collections of these molecules began manufacturing membranes to more effectively seal themselves off as a unit from other molecules in the world, and eventually collections of these protocells came together to form multicellular organisms, developing specialized sensory organs, motor organs, and the like. Presumably no one would deny that physicalism effectively captures everything that is happening in this world. At some point, the conglomerations of molecules start moving around the Life World in ways that are most efficiently characterized by adopting the intentional stance. We can get a snapshot of the phylogenic unfolding of this process by looking around at degrees of complexity in our current natural world. A protozoan senses nutrients and moves up the nutrient gradient because, we are tempted to say, it "wants" to eat. Schools of sardines swim tightly together for safety and initiate evasive action when presented with a predator-like stimulus. Small packs of savanna carnivores coordinate their movements in order to flush their prey out into the open where it can be more easily brought down. Chimpanzees engage in social deception, multistage planning for the future, and culturally transmitted tool use. Human beings read and write books about consciousness. Where and when in this chain of increasing complexity does ontological subjectivity pop into existence? Where does it *come* from?

Searle, of course, rejects mystical-substance dualism and has no patience for those who dogmatically declare that only humans have consciousness: when he comes home from work and observes his dog jumping about, wagging his tail, Searle is fairly confident that his dog is conscious, and indeed conscious of the specific emotional state of happiness. This confidence is based not simply on behavioral clues but because his dog is phylogenically closely related to himself and therefore the "causal underpinnings of [his] behavior are relatively similar to mine."[47] Searle is similarly confident that other close relatives of ours, such as chimpanzees, possess consciousness and notes that the presence or lack of consciousness in other animal species is an empirical issue, a matter of whether or not a given species has "a rich enough neurobiological capacity" to support consciousness.[48] Rich enough, however, does not seem like a particularly sharp distinguishing characteristic, whereas possession or lack of original intentionality apparently is. This gets to the heart of the problem with this concept from an evolutionary perspective:

whatever you call it—substance, property, X ontology—marking consciousness or intentionality off as a special something makes you a strong emergentist, and therefore essentially a dualist, whether you are comfortable with
those labels or not.

Back in the 1980s Searle tried to get across his sense that intentionality is
something special, as well as his conviction that the strong-AI project was
essentially doomed, with his famous Chinese Room thought experiment.[49]
Imagine a person in a room with all the Chinese characters in the language
at his disposal in the form of chits, as well as an incredibly detailed rule book
that tells him, when a certain sets of chits with characters is given to him
through a window (i.e., when he is given an input), what set of characters
to take down from the wall and slide back out in return (what to give as an
output). As Searle accurately predicts, our intuitive folk psychology tells us
that this person does not really understand Chinese: he is merely carrying
out a mindless algorithm. Searle takes this as proof that there is something
special about intentionality that cannot be captured by a mechanistic algorithm. Michael Arbib, however, draws the opposite and, I think, more
appropriate conclusion:

> [Upon considering Searle's Chinese Room argument] I am reminded of a clas-
> sic story about Norbert Wiener, the "father" of cybernetics. . . . Wiener believed
> he had resolved a famous nineteenth-century mathematical conjecture—the
> Riemann hypothesis—and mathematicians flocked from Harvard and MIT
> to see him present his proof. He soon filled blackboard after blackboard with
> Fourier series and Dirichlet integrals. But as time went by, he spent less and less
> time writing and more and more time pacing up and down, puffing at a black
> cheroot, until finally he stopped and said, "It's no good, it's no good. I've proved
> too much. I've proved there are no prime numbers." And this is my reaction to
> Searle. "It's no good, it's no good. He's proved too much. He's proved that even
> *we* cannot exhibit intelligence."[50]

In other words, what the Chinese Room thought experiment helps us get a
handle on is that, since there is no little person inside the Chinese Room that
is our brains, even *we* don't understand Chinese or English or anything—in
the strong sense of *understand* as *Verstehen*, or the unexplainable, mystical
workings of original or ontological Intentionality with a capital *I*.

Substance and property dualists aside, the final group of resisters to the
idea that mere physical processes can account for human consciousness
are what Patricia Churchland refers to as "boggled skeptics"[51]: the idea just
seems so damn hard to believe. John Locke expresses the boggled skeptics
position quite clearly: "For it is as impossible to conceive that ever bare in-
cogitative Matter should produce a thinking intelligent Being as that nothing

should of itself produce Matter."[52] This was, at one time, quite a powerful argument, despite its apparent simplicity: conscious beings seem to be able to do things that completely fly in the face of what we know about the behavior of inert matter. The conclusion that there has to be something else involved is therefore hard to avoid. As Dennett puts it, "Until fairly recently [the] idea of a rather magical extra ingredient was the only candidate for an explanation of consciousness that even seemed to make sense."[53] He goes on, however, to argue that the last few decades have seen the development of a crucial bit of evidence tipping things in favor of the physicalist view of consciousness: the development of artificial intelligence, which finally put to rest the boggled argument that no amount of physical complexity could produce consciousness-like phenomena. We have now built machines, which we know are just machines, that are capable of defeating Grand Masters at chess, passing the Turing Test (i.e., plausibly holding up their end of a free-form conversation), and demonstrating many of the powers that were previously seen as the exclusive province of conscious, intentional agents. Dennett observes that "the sheer existence of computers has provided an existence proof of undeniable influence: there are mechanisms—brute, unmysterious mechanisms operating according to routinely well-understood physical principles—that have many of the competences heretofore assigned only to minds."[54] As Hilary Putnam concludes, the overwhelming success of the physicalist model puts the folk model of dualism in an empirically untenable position, despite its intuitive appeal:

> We learn the so-called mental predicates by learning to use them in explanatory practices that involve embodied creatures. The idea that they refer to "entities" that might be present or absent independently of what goes on in our bodies and behavior has a long history and a powerful . . . appeal. Yet to say that the idea "might be true" is to suppose that a clear possibility has been described, even though no way of using the picture to describe an actual case has really been proposed.[55]

To say that soul-dependent theories "might be true" is thus a little generous; it is more accurate to say that they "appear to be false."

AI systems are still quite crude and extraordinarily inept at many tasks that are accomplished with ease by a five-year-old human. Similarly, there is still only a very rudimentary understanding of how the body-brain subserves even quite basic functions as memory, emotion, and self-consciousness Our current blind spots, however, should not be taken as proof that a useful and empirically rigorous science of human consciousness is a priori impossible. As Owen Flanagan notes, the current imperfect state of the field of the human-mind sciences often prompts a jump to mysterianism, and it is important to see how unnecessary and unjustified this jump is. He says,

> Although everyone thinks that cars and bodies obey the principles of causation—that for every event that happens there are causes operating at every junction—no one thinks that it is a deficiency that we don't know, nor can we teach, strict laws of automechanics or anatomy. . . . [So,] when an auto mechanic or a physician says that he just can't figure out what is causing some problem, he never says, "perhaps a miracle occurred."[56]

We might make a similar observation concerning the unpredictability of human thought and behavior, which is often cited as a sign of human beings' essential ineffability. It is exceedingly likely that, no matter how far the neuroscience of consciousness advances, it will remain impossible—because of sheer computational intractability, quantum randomness, whatever—to accurately predict the future behavior of even a single human being, let alone groups of human beings interacting with one another and with a constantly changing physical environment. It is equally likely that, no matter what advances we make in hydrology and meteorology, it will never be possible to pick out a single molecule of H_2O from the ocean inlet outside my window and predict where that molecule will be one year from now. We never for a minute, though, doubt that the molecule's future movements will be fully determined by the laws of physics. By extension, we have no more reason to believe that the cascades of neural impulses in our brains are any less determined and governed by physical causation than the water molecule.

Why should we want to block the move to ontological emergence, especially if it comes so naturally to us and takes so much work to get away from? Contrary to some doctrinaire physicalists, there is nothing about physicalism per se that makes it uniquely scientific. If we had an accumulation of a critical mass of replicable evidence for existence of some nonphysical, causally efficacious, intention-bearing substance, it would be unscientific *not* to be a dualist—and of course we cannot rule out the possibility that such point will ever be reached.[57] A pragmatic conception of scientific "truth" requires that our ideas of what could count as a viable explanation remain constantly open to revision. It just seems that physicalism is currently our best, most productive stance toward the world.

In the Game of Life example discussed above, we know that blinkers and gliders are not real because we have been shown how the system works: it is just a very large collection of squares turning ON and OFF. Despite what dogmatic materialists might say, we do not know with the same kind of certainty that there is no such thing as a soul. We don't know *anything* about our world for sure, because—unlike the artificial Life World—we were not invited to observe its creation. What we *can* say, though, is that the best evidence for the existence of a soul has been undermined by AI, which is, in essence, an exceedingly crude, Life World–like simulation of a person. In addition, everything that we know about how the world in general works suggests that there is no place for nonphysical causation. No one has come up with a story that would explain how something like a soul could exist in the world as we currently understand it, although we are, qua human beings, highly motivated to come up with and to believe such stories. Given the explanatory track record of physicalist accounts of the universe, as well as the failure thus far of highly motivated humans to come up with an empirically viable alternative to it,[58] it would thus seem that a strong emergentist view of mental or intentional properties can only be defended as an article of religious faith. In the absence of an empirically defensible account of dualism, the explanation of reality that best enables us to get a grip on the world does not involve ghosts, souls, miracles, or original intentionality: human beings, like all of the other entities that we know about, appear to be robots or zombies all the way down, whether we like that idea or not.

The Limits of Physicalism: Why We Will Always Be Humanists

Having hopefully blocked the move to mysterianism or ontological emergentism, I would here like to address in more detail the issue of why these intellectual moves are so compelling to us, as well as what this compulsion *does* reveal about the special status of human-level concepts. I argued above that John Searle is engaging in a bit of philosophical sleight-of-hand when he purports to be a biological materialist but then continues to insist on a special ontological status for human subjectivity. Searle is a brilliant philosopher with a quite-detailed grasp of the state of the field in the cognitive and neurosciences. Why, then, this refusal to relinquish the idea of two distinct ontologies? And why two, we might ask, and not three? Or ten? In this section I explore the intuition that I think motivates the defenders of dualism in all of its various forms: the recognition that human-level reality is real for humans and that it is so deeply entrenched that no third-person description can ever completely dislodge it. In other words, we apparently cannot help but at some level see a *Geist* in the machine, which means there will always be something importantly different about the *Geisteswissenschaften*.

Why Physicalism Doesn't Matter

Hard-core physicalists such as Dennett are inclined to dismiss positions such as Searle's or Nagel's as mere statements of religious belief or personal sentiment. Dennett and some other advocates of vertical integration argue that, since intentionality and consciousness are like gliders in the Game of Life—helpful for certain heuristic purposes but with no underlying reality—the rigorous study of human affairs will eventually be able to dispense with them entirely. Owen Flanagan, for instance, urges us to get beyond such concepts as the soul or free will: "Since these concepts don't refer to anything real, we are best off without them."[59] He compares such dualistic concepts to shadows on the wall of Plato's cave—the result of mistaking appearance for reality.[60] Flanagan also suggests that our belief in the soul is the result of our particular cultural tradition, traceable back to the Judeo-Christian worldview, and is therefore best seen as a passing fashion.[61]

A common analogy drawn by those who feel dualism will soon go the way of bell bottoms and disco balls is the shift in human sensibilities that occurred with the Copernican revolution. Copernicanism presented a view of the solar system that contradicted not only scriptural authority but the evidence of our senses: the Bible states quite clearly that the sun moves around the Earth, and this also happens to accord with our everyday sensory experience. Yet the accumulation of empirical evidence eventually resulted in Copernicanism winning the day—trumping both religion and common sense—and nowadays every educated person takes the heliocentric solar system for granted. Dennett argues that the current physicalism-versus-dualism controversy is analogous to the early days of Copernicanism: we are resistant to physicalism because it goes against our religious beliefs and our common sense, but the weight of the empirical evidence is on its side. Eventually—after all of the controversy has played itself out—we will learn to accept the materialist account of the self with as much equanimity as the fact that the Earth goes around the sun.[62] Dennett illustrates our tendency to seek Intelligence with a capital *I* in the universe with a song he learned as a child titled "Tell Me Why":

> Tell my why the stars do shine,
> Tell me why the ivy twines,
> Tell me why the sky's so blue;
> Then I will tell you just why I love you,
> Because God made the stars to shine.
> Because God made the ivy twine.
> Because God made the sky so blue.
> Because God made you, that's why I love you.[63]

Just as children come eventually to realize that there is no such thing as Santa Claus or the tooth fairy, so will dualistic adults eventually grow up intellectually, realize the truth of physicalism, and come to see such songs and sentiments as childish wishful thinking—beautiful and once quite comforting, but not a proper component of a mature worldview.

A basic problem with this position, however, is that there is a profound disanalogy between the Copernican revolution and the revolution represented by physicalist models of the mind. The Ptolemaic model of the solar system falls quite naturally out of the functioning of our built-in perceptual systems, but it is not itself part of that system: we do not appear to possess an innate Ptolemaic solar-system module. Switching to Copernicanism, at least intellectually, requires us to suspend our commonsense perceptions, but it does not involve a direct violation of any fundamental, innate human ideas. Physicalism as applied to mind *does* require such a violation, and this has a very important bearing on how realistic it is to think that we can dispense with mentalistic talk once and for all. Owen Flanagan characterizes dualism as something that has troubled us "for centuries."[64] But seeing agents as something special goes back for at least as long as people have had theory of mind—perhaps one hundred thousand years.[65] This is the psychological fact behind the argument forwarded by Searle and others that consciousness is special: it is inescapably real for us.

We Are Robots Designed Not to Believe That We Are Robots

The idea of human beings as ultimately mindless robots, blindly designed by a consortium of genes to propagate themselves, has so much difficulty gaining a foothold in human brains because it dramatically contradicts other factory-issued and firmly entrenched ideas such as the belief in *soul, freedom, choice, responsibility*—the products of our theory of mind. Human beings appear to be born dualists.[66] That is, our innate theory of mind causes us to feel that there is something special about human agents and that this specialness has to do with a kind of mental or spiritual substance completely distinct from the physical world and its laws. The dualism advocated by Plato and Descartes was not a historical or philosophical accident but rather a development of an intuition that comes naturally to us: agents are different from things. Agents actively think, choose, and move themselves, whereas things can only be passively moved. The locus of agents' ability to think and choose is the mind, and because of its special powers the mind has to be a fundamentally different sort of entity than the body. Even cultures that did not develop a doctrine of strong mind-body substance dualism—such as the early Chinese—nonetheless believed that there was something special about the mind. As the fourth-century B.C.E. Chinese thinker Mencius put it, what distinguishes the heart-mind (*xin*)—the locus of agency in human beings—from other organs of the body is that it issues commands, whereas the other parts of the body merely follow them.[67] The idea that human beings, like all other animals on the planet, are physical systems produced by a purposeless process of differential reproduction combined with natural selection is thus fundamentally counterintuitive to us because it denies that we have minds, which in turn seems to contradict our conviction that we possess free will and the dignity and responsibility that goes along with such autonomy.

We cannot help but see the physical movements of other human beings as caused by the free workings of a nonmaterial mind, full of desires, thoughts, beliefs, and fears. Our theory of mind also seems to be somewhat overactive. We have a tendency to project these nonmaterial mental qualities onto almost *anything* that moves in a particular kind of way. Geometric shapes in a short animation or single dots moving around on a screen appear irresistibly to us to be involved in goal-directed, mentalistic behavior and for this reason engage our sympathy. In his 1993 book, *Faces in the Clouds: A New Theory of Religion*, Stewart Guthrie argues that our tendency to overproject agency onto the world is what gives rise to religion—in his definition, the belief in supernatural beings.[68] He portrays this universal tendency to see faces in the clouds and rocks as bears as the evolutionary equivalent of Pascal's wager. Pascal, of course, famously argued that we can never be sure that God does not exist; considering that the cost of believing in God is finite, while the consequence of not believing in Him, if He does exist, is infinite (eternal damnation), it would be irrational not be a theist. Guthrie argues that overprojection of agency is the result of a similar, evolutionary wager:

> We animate and anthropomorphize, because, when we see something as alive or humanlike, we can take precautions. If we see it as alive we can, for example, stalk it or flee. If we see it as humanlike, we can try to establish a social relationship. If it turns out not to be alive or humanlike, we usually lose little by having thought it was. This practice thus yields more in occasional big successes than it costs in frequent little failures. In short, animism and anthropomorphism stem from the principle "better safe than sorry."[69]

Although Guthrie's book makes no reference to the innate cognitive modules dedicated to theory of mind and animate objects, it is easy to see how the phenomena he discusses can be understood as the result of the overapplication of these modules—their projection onto domains for which, strictly speaking, they were not originally designed.

Guthrie does an exhaustive job of documenting the long history and pervasive presence of anthromorphism and animism in human perception, art, philosophy, science, and religion. It is clear that human beings, no matter how professionally or intellectually committed to physicalism, feel a constant compulsion to project agency onto the inanimate. We all know this experience, having to deal daily with stubborn, diabolical computers bent on erasing our data, crotchety old cars that refuse to start, and beloved old pairs of pants that finally have to be laid to rest—or, as in my case, that have to be carefully kept hidden from one's wife in order to avoid their being forcibly put out of their misery. The anthropomorphic drive seems to be universal and appears quite early in development. Deborah Kelemen has documented the widespread projection of invisible or supernatural agency onto the world[70]—what she refers to as promiscuous *teleology*—in children of various ages and education levels and argues that agent-centered, teleological explanations for phenomena seem to be the human cognitive-default position, only gradually, with difficulty, and incompletely dislodged by mechanistic explanation.[71]

We are obviously capable of withdrawing our projections when we have to—recognizing that the beloved pants are just pieces of fabric or that our computer is not really out to get us—but it takes cognitive effort, which suggests that it does not come naturally and is not easily sustainable. As Steven Pinker notes, this is why even researchers in the cognitive sciences and neurosciences—of all people, the ones who should be clearest about the absolutely physical nature of the human mind—usually end up smuggling some form of dualism back into their picture of the human being by means of what he calls the Pronoun in the Machine, a vague seat of free will and human dignity, stripped of metaphysical overtones but essentially fulfilling the same function as Descartes' ghost. "As men of scientific acumen," Pinker observes, "they cannot but endorse the claims of biology, yet as political men they cannot accept the discouraging rider to those claims, namely that human nature differs only in degree of complexity from clockwork."[72]

The power of human cognitive defaults should not be underestimated. In religious studies, we see the power of folk theory manifest itself in the gap between certain religious doctrines and the way people process these doctrines and apply them to their lives. The Buddhist doctrine of *anatman* (no self), for instance, is about as counterintuitive as the modern neuroscientific model of the mind. Indeed, as many scholars have argued, there seems to be a fair amount of overlap between the two.[73] Perusal of the Pali Canon—the earliest records of the Buddha's teachings—reveals the immense amount of difficulty that the historical Buddha, Siddhartha Gautama, apparently had in trying to get his disciples to grasp the concept of no self.[74] It is also very revealing that, not long after the death of the historical Buddha, the various descendent strands of Buddhism that eventually labeled themselves as the Great Vehicle (*Mahayana*) essentially snuck the self back in through such doctrines as Buddha nature and that, once Buddhism spread to East Asia, this understanding of Buddha nature took on an obviously substantial form. Although to this day there are Buddhists that advocate and defend the orthodox no-self doctrine, and despite the centrality of no self to Buddhist teachings, the vast majority of Buddhists throughout history seem to have been unable or unwilling to seriously entertain it.[75]

The pervasive, subtle power of innate modules appears to contaminate every attempt to break away from ordinary human thought. Consider the example of Albert Camus and his vision in the *Myth of Sisyphus* (1942) of *l'homme absurde*, who supposedly sees the world as it appears through the lens of physicalism: mechanistic, unfeeling, and meaningless. I have essentially argued above that Camus is right about this much: we *do* live in a mechanistic, meaningless universe. Yet we are mistaken if we think that insight into lower levels of causation can, in any existential sense, completely free us from the higher-level structures of meaning in which we are innately entwined. Despite its surface bleakness, Camus' vision strikes many people—including myself—as powerful and beautiful. Why is this? It is because, despite Camus' conceit that he has freed himself from false consciousness, works like *The Myth of Sisyphus* are inextricably permeated with human-level values such as clarity, freedom, and strength and the fundamental motivation of such work is the wonderful feeling of control and understanding that we acquire when we have seen through surface appearances to the very truth of things. Camus' creativity consists in recruiting these innate, universal normative reactions and mapping them in a quite novel manner—lucidity consists in knowing nothing for certain, and courage consists in rejecting those transcendent truths that once were perceived as requiring strength to defend against unbelief—but the sources are probably as old as *Homo erectus.*

It is thus a mistake to say that we will ever completely dispense with mentalistic concepts or ever entirely succeed in withdrawing our projections from the world. My disagreement in this regard with, for instance, Dennett, is best illustrated by the respective songs we would choose to illustrate the human tendency to project theory of mind onto the world. Dennett's choice is a children's song, one that we now look back at with affection but also with a hint of indulgent superiority—it was a comforting thought, but we know better now. I think that a much more representative choice would be Nina Simone's rendering of the song "Feeling Good,"[76] which rhapsodizes about the joy that we feel when seeing lazy reeds drifting by on a stream, rivers running free, and butterflies enjoying themselves in the sun. It is a rare cognitively intact person who can listen to Simone without feeling in their bones the emotional and mental contagion that is constantly taking place between human beings and their world. The sight of "reeds driftin' on by" can make us feel calm, and a feeling of calmness can color our perception of the reeds. Rivers really do seem to run free, and the play of butterflies cannot help but seem fun to us—even though, qua scientists, we know at some level that nothing is really going on except water molecules being drawn downward by gravity and some large insects engaged in a random feeding pattern. Most importantly of all, feeling this kind of resonance between our own concerns and the functioning of the universe makes us feel really, really *good*.

This feeling is, I think, why Pascal's wager, Guthrie's theory of agency overprojection, is incomplete as well as why Dennett is wrong about our potential to completely leave dualism behind. Our promiscuous teleology and overactive theory of mind play a less-accidental and less-peripheral role in the economy of the human psyche than the simple better-safe-than-sorry explanation would have it. As the basis of perceiving meaning in the world, theory of mind would appear to be the foundation of any kind of long-term, large-scale motivation. I can be moved to engage in short-term, limited acts—consuming a cheeseburger when hungry or seeking out sleep when tired—without inquiring into the meaning of what I am doing, but the universal and pervasive tendency of human beings to tell and hear stories answering the question *why* suggests that long-term planning and motivation requires such a sense. The feeling that our work or our life has a purpose involves embedding it in an at least implicit narrative, and the agent-centered nature of such narratives suggests to me that the human ability to remain motivated over the course of long-term, multistep, delayed-gratification tasks involves the evolutionary hijacking of reward centers in the brain whose original or proper domain is interpersonal approval and acceptance. In cognitively fluid humans, reward expectancy over long-term tasks may be maintained at least in part by the feeling that some metaphorical conspecific "up there" is watching and approving or disapproving of our actions or (in its modern iteration) in a more diffuse, nontheistic sense that what we are doing matters—a conceit that makes no sense unless we project some sort of abstract, metaphorical agency onto the universe. I would also suggest that in suicidal depressives we see a breakdown of this system: severely depressed individuals are actual realizations of Camus' *homme absurde* and genuinely *do* seem to perceive the world as unfeeling, mechanistic, and meaningless all the way down. The result is not a feeling of clarity or power, however, but profound behavioral paralysis and overwhelming suicidal tendencies.[77] Evolution is a tinker, and when faced with the task of getting live-in-the-moment social animals to start thinking in more complex and indirect ways about the long term, it simply coopted a previously existing and very big carrot and stick. Prehuman social animals are powerfully motivated to shape their behavior in such as way as to win the approval and avoid the approbation of their literal social group. The great cognitive innovation that led to human beings—cognitive fluidity, the ability to project from one domain to another—perhaps also enabled literal social approval and disapproval to be projected onto a much larger scale: not just our immediate tribe, but the cosmos itself.

We will apparently always see meaning in our actions—populating our world with "angry" seas, "welcoming" harbors, and other human beings as unique agents worthy of respect and dignity and distinct from objects in some way that is hard to explain in the absence of soul-talk but nonetheless very real for us. We will continue to perceive our work, families, and lives as being meaningful on some inchoate level and to be strongly motivated to make the appropriate changes whenever we begin to lose this sense.

Qua scientists, we can acknowledge that this feeling is, in some sense, an illusion. For better or worse, though, we are apparently designed to be irresistibly vulnerable to this illusion—in this respect, appearance *is* reality for us human beings. This is where, in fact, we see the limits of a thoroughly scientific approach to human culture and need to finesse a bit our understanding of what counts as a fact for beings like us.

Human Reality Is Real

Humanists and natural scientists concerned with the issue of levels of explanation and emergent properties have much to learn from the work of the Canadian philosopher Charles Taylor. Taylor is a humanist who has grappled with vertical integration and come away unimpressed, and he sees his work as a defense of humanism against the reductionistic threat posed specifically by sociobiology and more generally by the broader naturalistic bent of the modern world. We do not have to follow Taylor to his conclusion, which is essentially to reaffirm the Cartesian gulf between the *Geistes-* and *Naturwissenschaften*, in order to feel the power of his basic position. His conception of human-level reality provides us with a nuanced, sophisticated model for understanding the place of the person in the great physicalist chain of causation.

One of Taylor's most important points is that human beings, by their very nature, can only operate within the context of a normative space defined by a framework of empirically unverifiable beliefs. The Enlightenment conceit that one can dispense with belief or faith entirely and make one's way through life guided solely by the dictates of objective reason, is nothing more than that: a conceit, itself a type of faith in the power of a mysterious faculty, reason, to reveal incorrigible truth. In addition to the panoply of "weak evaluations"— such as a preference for chocolate over vanilla ice cream—that we are familiar with, humans are also inevitably moved to assert "strong" or normative evaluations. This latter type of evaluation is based on one or more explicit or implicit ontological claims and therefore is perceived as having objective force rather than being a merely subjective whim. For instance, I don't particularly like chocolate ice cream and believe that the flavor of vanilla ice cream is superior. I don't, however, expect everyone to share my preference and am certainly not moved to condemn my wife for preferring chocolate. I am also not inclined to sexually abuse small children, but this feels like a different sort of preference to me: abusing small children seems *wrong*, and I would condemn and be moved to punish anyone who acted in a manner that violated this feeling. If I were pressed on the matter, this condemnation would be framed, moreover, in terms of beliefs about the value of undamaged human personhood and the need to prevent suffering and to safeguard innocence.

All of the classic Enlightenment values that we continue to embrace as modern liberals—the belief in human rights, the valuation of freedom and creativity, the condemnation of inflicting suffering on innocents—are strong evaluations of this sort, dependent on an implicit set of beliefs about human beings historically derived from Christianity but reflecting common human normative judgments. Although the Enlightenment *philosophes* began disengaging these beliefs from their explicitly religious context, and we in the last century have more or less completed this process, this does not change their status as beliefs. The "self-evident truths" enshrined in such classic liberal documents as the Declaration of Independence of the United States and the UN's Universal Declaration of Human Rights are not revealed to us by the objective functioning of our a priori reason but are, rather, items of faith.

Taylor argues that metaphysically grounded normative reactions such as these are inevitable for human beings. The fact that we cannot coherently account for our own or other's behavior without making reference to them—as well as the fact that they irresistibly present themselves to us as objective despite our lack of proof for them—says something important about what it means for something to be real for human beings. Although values are not part of the world as studied by natural science, the fact that value terms such as *freedom* and *dignity* are "ineradicable in first-person, nonexplanatory uses"[78] means that they are, in a nontrivial sense, real. "[Human] reality is, of course, dependent upon us, in the sense that a condition for its existence is our existence," Taylor concedes. "But once granted that we exist, it is no more a subjective projection than what physics deals with."[79] For the peculiar type of animal that we are, moral space is as much a part of reality as physical space, in that we cannot avoid having to orient ourselves with respect to it.

To reformulate Taylor's insights into the naturalistic framework I have been arguing for here, we can say, qua naturalist, that our overactive theory of mind causes us to inevitably project intentionality onto the world—to see our moral emotions and desires writ large in the cosmos. It would be empirically unjustified to take this projection as real. Nonetheless, the very inevitability of this projection means that, whatever we may assert qua naturalists, we cannot escape from the lived reality of moral space. As neuroscientists, we might believe that the brain is a deterministic, physical system like everything else in the universe and recognize that the weight of empirical evidence suggests that free will is a cognitive illusion.[80] Nonetheless, no cognitively undamaged human being can help *acting* like, and at some level really *feeling*, that he or she is free.[81] There may well be individuals who lack this sense, and who can quite easily and thoroughly conceive of themselves and other people in purely instrumental, mechanistic terms, but we label such people "psychopaths," and quite rightly try to identify them and put them away somewhere to protect the rest of us.[82] Similarly, from the perspective of evolutionary psychology, I can believe that the love that I feel toward my child and my relatives is an emotion installed in me by my genes in accordance with Hamilton's Rule. This, however, neither makes my experience of the emotion, nor my sense of its normative reality, any less real to me. Indeed, this is precisely what I would expect from the third-person perspective: the gene-level, ultimate causation wouldn't *work* unless we were thoroughly sincere at the proximate level. The whole purpose of the evolution of social emotions is to make sure that these false feelings seem inescapably real to us, and this lived reality will never change unless we turn into completely different types of organisms. As Steven Pinker observes concerning our moral intuitions, "Whatever its ontological status may be, a moral sense is part of the standard equipment of the human mind. It's the only mind we've got, and we have no choice but to take its intuitions seriously. If we are so constituted that we cannot help but think in moral terms (at least some of the time and toward some people), then morality is as real for us as if it were decreed by the Almighty or written into the cosmos."[83]

This principle is not confined to natural scientific explanations of human phenomena: it holds with regard to *any* sort of objective, reductive, third-person explanation, even those coming from within the humanities, such as Taylor's genealogy of the construction of the modern Western self. For instance, qua historian, I believe that my sense that human rights ought to be defended is a contingent item of faith, a product of the Enlightenment that I happen to have inherited because of the age and place in which I was born. This abstract insight is potentially quite helpful in facilitating communication with people from other moral traditions, but such interactions are unlikely to shake my faith in the importance of, say, women's rights—unless this faith is, perhaps through conversion to a fundamentalist strand of traditional religion, replaced by an opposing belief. Completely extracting ourselves from moral space is as impossible as stopping our visual systems from processing information when we open our eyes or our stomach from registering distress when our blood-sugar level drops below a certain point. In this sense, then, human-level truth is inescapably real.

The Importance of Physicalism:
Why Physicalism Both Does and Does Not Matter

To the extent that human-level reality will always have a hold on us, then, we are entitled to say that physicalism does not matter. This leads Taylor to conclude that the unavoidability of human-level concepts is not merely a phenomenological observation but rather a clue as to the "transcendental conditions"[84] of "undamaged human personhood,"[85] and thereby a refutation of any sort of third-person, naturalistic account of the humanities.[86] If human reality is indeed real for us, why *not* follow Taylor and say that it is just as real as anything studied by the natural sciences?

Why Physicalism Does Matter

In short, because it is not. Or, to put this more accurately, our innate empirical prejudice appears to be so constituted that, once we have *explained* something—that is, reduced a higher-level phenomenon to lower-level causes—the higher-level phenomenon inevitably loses some of its hold on us.[87] Despite what Stephen Jay Gould and the AAAS say,[88] there is an important difference between literally believing that God created the world in seven days and thinking that this is a beautiful story that can mean something to us on Sundays but must be put aside when we go about our daily work. Evolution is such a relatively new idea, and its message so fundamentally alien to us, that its real implications for our picture of human reality have yet to fully sink in, which is why most liberal intellectuals continue to believe that Darwinism does not seriously threaten traditional religious beliefs or conceptions of the self. It clearly *does*, however, and once we have begun down the physicalist path we cannot go back to the old certainties. This is not merely because it would be illogical to do so—although it would—but because we just seem to be built in such a way that we want to deal with and picture the world as it "really" is, no matter how unpleasant.

We can get a sense of this human "truth" prejudice—really, a preference for lower-level over higher-level explanation—by thinking about a typical reaction to a science-fiction movie that was very popular some years back called *The Matrix* (1999). For those unfamiliar with the plot, the protagonist,

Neo, begins to uncover puzzling clues that his everyday world is an illusion. He eventually discovers that his body and those of others in his apparently real world of the Matrix are, in fact, being maintained in sinister life-support tanks housed in a vast factory. Their brain activity is being farmed as a source of energy by the evil machines who created the Matrix—an elaborate virtual world, projected onto the brains of the bodies in the tanks—in order to fool their prisoners into thinking that they are free. Neo eventually gets in touch with a doughty band of humans who have liberated themselves from the life-support tanks and who live crude, uncomfortable, but free lives in an underground refuge called (rather heavy-handedly) Zion.

One of the more interesting points in the movie is when a cowardly informant is induced to betray the inhabitants of Zion and return to the tanks in exchange for a particularly pleasant illusory life-style: in the virtual world of the Matrix, he is to be a rich and powerful man, with every sensory pleasure one could desire. Most importantly, he will not *remember* that this is all an illusion: his fine steak and excellent red wine will taste just as good as the "real" things would, and the pleasure he will derive from his new virtual life will be—to him, at least—inescapably true and powerfully felt. Especially when compared to the threadbare and uncomfortable life in the bleak underground burrow of Zion, this seems like a pretty good deal: If you don't know the Matrix is not real, what difference does it make? If the steak tastes like steak, why should you care that you are really pickled in a tank and being farmed by evil machines? If your memories are to be perfectly erased, why would it matter that you had betrayed your comrades and your former cause?

It probably *wouldn't* matter. The important thing, though, is that we, as human beings, feel that it *would*—we feel anger with this traitor, as well as revulsion at the idea of returning voluntarily to the Matrix. Why? Because, as Aristotle said, we are constituted in such as way as to desire the Good, and the Good for human beings involves being properly situated with regard to what we feel to be the truth. Promised future rewards that we know to be illusory seem less valuable to us, even if we are assured that they will *seem* real when we get them. The same inchoate instinct that makes life in the Matrix abhorrent to us makes it impossible to continue to embrace, at least in precisely the same way, traditional religious ideals that appear to be in conflict with what we are convinced we now know about the world. And—at least as long as physicalism remains our current best explanation of the world—any religious or philosophical belief based on dualism is going to be in this sort of conflict.

This is where the Copernican analogy *is* helpful. We quite happily live our everyday lives in a Ptolemaic solar system, seeing the sun rise and set and enjoying the felt stability of the earth under our feet. We acknowledge, though, that this appearance is an illusion, and that the Earth is really racing through space at 108,000 kilometers per hour around the sun. Why does it matter what is really the case if it makes no difference to the way we see things? It matters because making important, practical decisions based on what is really the case, as opposed to what seems to be the case, works better. Launching satellites or sending off space probes simply would not work very well unless we suspended our intuitive Ptolemaic worldview when engaged in this sort of work. The same is true of human-level realities. The realization that the body-mind is an integrated system is counterintuitive, but treatments based on this insight appear to be massively more effective than dualism-based treatments—pharmaceutical interventions, for instance, have done more for the treatment of mental illness in a few decades than millennia of spiritual interventions, from exorcisms to Freudian analysis. Recognizing that there is no point at which the ghost enters the machine allows us to go ahead with stem-cell research, and understanding that personhood is not an all-or-nothing affair helps us get a better grip on what is going on with severe dementia in the elderly. Physicalism matters because it simply works better than dualism, and—once the reality of this superiority is fully grasped—this pragmatic consideration is an irresistibly powerful argument for creatures like us.

Dual Consciousness: Walking the Two Paths

How can physicalism both matter and not matter? We can answer this question by returning to Nietzsche's critique of Kant and seeing how *both* Niezsche and Kant were right. To take moral intuitions as an example, we can follow Nietzsche—somewhat updated and put into the role of an evolutionary psychologist—and see why it is important and revealing to ask about the adaptive forces that cause us to feel the force of synthetic a priori claims, rather than simply experiencing them as unquestioned intuitions. Answering the question of origins—uncovering the lower-level, ultimate explanations for our moral intuitions—has important practical implications, but most of all we just simply want to *know*. We also need to follow Kant, however, in recognizing that, no matter what the origins of these intuitions, they are the spontaneous product of a very powerful, built-in faculty, the output of which seem inescapably right to us. This means that, as an empirical responsible humanists—or even humans *tout court*, living in the modern world of pharmaceuticals and behavioral neuroscience—we need to pull off the trick of simultaneously seeing the world as Nietzsche and as Kant, holding *both* perspectives in mind and employing each when appropriate. Those who have allowed the acid of Darwinism to finally breach the mind-body barrier thus end up living with a kind of dual consciousness, cultivating the ability to view human beings simultaneously under two descriptions: as physical systems and as persons. On the one hand, we are convinced that Darwinism is the best account we have got for explaining the world around us, that, therefore, human beings are physical systems, and that the idea that there is a ghost in the machine should be abandoned. On the other hand, cognitively intact humans apparently cannot help but feel the strong pull of human-level truth.

Taking the neuroscientific model of the self seriously thus involves a balancing act that serves as a testament to our human ability to hold multiple, mutually contradictory perspectives in our minds at once. This basic adaptive function of this ability was probably originally, and continues to be, the ability to hold multiple *social* perspectives at once, which is a crucial talent for functioning successfully in even a moderately sized social group. Among other things, it is crucial for social deception: when you lie, you need to keep at least two different models of the world in your mind at once, and we humans are remarkably good at this exceedingly difficult task. This ability to simultaneously hold different perspectives in mind, though originally evolved for the purpose of social deception, can apparently be exapted and applied to explanations of the world.[89] It forms the basis of our ability to entertain alternative perspectives, to see anything in the world qua some role—qua physicialist or qua everyday person reading a novel.

We should also recognize that such mental spaces, by their very nature, are not all created equal—otherwise they would be indistinguishable from each other. One of the most difficult things about lying, for instance, is that we appear to be unable to maintain as detailed and vivid a model of the lie-world as our real world. This leads to certain problems—we can be tripped up by our lies in a way that is impossible if we are telling the truth—but it is probably a crucial design feature. For if both spaces were equally powerfully realized, we would lose track of the lie, which would defeat the very purpose of lying. The same principle holds with mentally visualized imagery versus real-time perception: an imagined object or scene is always less vivid that the real thing, and it is extremely important that this be so, lest we become confused about whether we were really being chased by a tiger or just worrying about the possibility.

I would speculate that, when it comes to dual consciousness regarding physical and human levels of explanation, the human level will always be much more vivid and real to us. Evolution would have done a poor job if it were otherwise, and evolution does not do poor jobs. This means that taking the physicalist stance will require the same sort of effort and training that any counterintuitive ability requires, and our ability to remain in the physicalist mental space will always be somewhat limited. Francis Crick, at one point in his discussion of the astonishing hypothesis that the mind is nothing other than the brain, observes somewhat bemusedly, "I myself find it difficult at times to avoid the idea of a homunculus. One slips into it so easily."[90] Crick was not alone. The interpretive space involving the perception of agency and consciousness, in ourselves as well as others, is a product of one of the most basic of human cognitive modules and we will never escape its gravitational pull for long. The fact that Richard Dawkins and Daniel Dennett remain at large, simultaneously espousing thorough-going physicalism in their professional work while—at least to the best of my knowledge—continuing to participate in and derive pleasure from normal human relationships and goal-directed activities, says more about the robustness and automaticity of our theory of mind space than the plausibility of fully and continuously embracing a mechanistic view of the universe.

We can mine the world's religious traditions for helpful metaphors for what this kind of dual stance toward the world might be like. Jesus famously advised his followers to be "in the world but not of it,"[91] and the figure of the Bodhisattva in the Mahayana Buddhist tradition dwells simultaneously in two realms: that of ultimate truth, where there are no distinctions and no suffering and therefore no need for compassion or Buddhism, and that of conventional truth, where suffering is real and the Bodhisattva is called upon to exercise finely tuned and deeply felt compassion. My favorite analogy comes from the fourth century B.C.E. Chinese thinker Zhuangzi, who describes his ideal person as "walking the two paths"—that of the heavenly/natural (*tian*) and the human. From the heavenly/natural perspective, there are no distinctions, no right and wrong, no feelings, no truth. From the human perspective, all of these things are acutely real. The key to moving successfully through the world, Zhuangzi believes, is simultaneously keeping both perspectives in mind, seeing the human "in the light of the heavenly," and thus seeing through to its contingent nature, while at the same time acting in accordance with the constraints of being a human in the world of humans.[92] This kind of dual consciousness is also perhaps what Kant was getting at in a curious passage from the *Groundwork* when he declares that we must "lend" the idea of freedom to rational beings:

> Now I assert that every being who cannot act except under the Idea of freedom is by this alone—from a practical point of view—really free; that is to say, for him all the laws inseparably bound up with freedom are valid just as much as if his will could be pronounced free in itself on grounds valid for theoretical philosophy. And I maintain that to every rational being possessed of a will we must also lend [*leihen*] the Idea of freedom as the only one under which he can act.[93]

We know, qua physicalist, that we are not free, but in our everyday lives we cannot help acting as if we are free, lest we find ourselves exiled from the Kingdom of Ends—that is, no longer recognizable as undamaged human agents.

Embracing the Neuron without Losing the Human

To conclude, then, we should not allow our distaste for physicalist explanations of the human to turn us into reactionaries. The subject of humanist
inquiry is not the workings of some Cartesian *Geist* in the machine, but rather
the wonderfully complex set of emergent realities that constitute the lived
human world, in all its cultural and historical diversity. The realization of the
thoroughly physical nature of this reality does not condemn us, however, to
live forever after in an ugly world of things. For undamaged humans, other
humans can never be a existentially grasped as mere things,[94] and our promiscuous projection of teleology onto the world assures that we will continue to
find the whole materialist universe a rather beautiful place once it is properly
understood. The fact that even the most resolutely physicalist conception of
the world cannot help but continue to inspire awe and an implicit sense of
meaning in human beings is captured quite well in the character of Henry
Perowne, a neurosurgeon and committed materialist, in Ian McEwan's recent
novel *Saturday*. Prompted by a poem to imagine being "called in" to create a
new religion, Perowne declares that he would base his upon evolution:

> What better creation myth? An unimaginable sweep of time, numberless genera-
> tions spawning by infinitesimal steps complex living beauty out of inert matter,
> driven on by the blind furies of random mutation, natural selection and envi-
> ronmental change, with the tragedy of forms continually dying, and lately the
> wonder of minds emerging and with them morality, love, art, cities—and the
> unprecedented bonus of this story happening to be demonstrably true.[95]

Our innate cognitive mechanisms ensure that the modern scientific model of
human beings as essentially very complicated things will not lead to nihilism
or despair. In the end, acknowledging our inescapable embodiment not only
possesses the excellent advantage of being demonstrably true but also cannot
help but enrich our sense of wonder at the dependent and tragic human condition, detracting nothing from the felt beauty and nobility of it.

Notes

1. Stephen Jay Gould, "Nonoverlapping Magisterial," *Natural History* 106 (1997):
16–22. Cf. Michael Ruse, *Can a Darwinian Be a Christian? The Relationship between
Science and Religion* (New York: Cambridge University Press, 2001). This is also the
position advanced by the American Association for the Advancement of Science—the
world's largest scientific society—in their recent and widely publicized book on the
evolution-creationism debate: Catherine Baker, *The Evolution Dialogues: Science,
Christianity, and the Quest for Understanding*, ed. James B. Miller (Washington, D.C.:

Program of Dialogue on Science, Ethics, and Religion / American Association for the Advancement of Science, 2006).

2. Francis Crick, *The Astonishing Hypothesis: The Scientific Search for the Soul* (New York: Simon & Schuster, 1994), 259.

3. "Sì, abbiamo un'anima. Ma è fatta di tanti piccoli robot." From Giulio Giorelli, "Sì, abbiamo un'anima. Ma è fatta di tanti piccoli robot" (interview with Daniel Dennett), *Corriere della Sera* (Milan), April 28, 1997, quoted in Daniel Dennett, *Freedom Evolves* (New York: Viking, 2003).

4. Steven Pinker, *The Blank Slate: The Modern Denial of Human Nature* (New York: Viking, 2002), 72.

5. Edward Slingerland, *Effortless Action: Wu-Wei as Conceptual Metaphor and Spiritual Ideal in Early China* (New York: Oxford University Press, 2003).

6. Although if I *had* done this as an academic exercise, it would have to have been written as an imitation transparent enough (note the metaphor) to serve as a commentary on how such texts were structured.

7. Gottfried Wilhelm Leibniz, *New Essays on Human Understanding*, trans. and ed. Peter Remnant and Jonathan Bennett (New York: Cambridge University Press, 1996), 68.

8. Literally "by means of a means."

9. Friedrich Nietzsche, *Beyond Good and Evil*, trans. Walter Kaufmann (New York: Vintage, 1966), 18–19.

10. Nietzsche's wonderful term for this sort of circular or nonsensical reasoning, after the Molière play from which the doctor's speech is cited.

11. John Tooby and Leda Cosmides, "Psychological Foundations of Culture," in *The Adapted Mind: Evolutionary Psychology and the Generation of Culture*, ed. Jerome Barkow, Leda Cosmides, and John Tooby (New York: Oxford University Press, 1992), 19–136.

12. See Daniel Dennett, *Darwin's Dangerous Idea: Evolution and the Meaning of Life* (New York: Simon & Schuster, 1995), 166–81. The Game of Life was first presented to a wide audience in Martin Gardner, "Mathematical Games," *Scientific American* 223 (1970): 120–23; also in Martin Gardner, "Mathematical Games," in *Scientific American* 224 (1971): 112–17. It has since become an Internet phenomenon, with innumerable sites dedicated to working out Life Worlds with various rules and initial states.

13. Gardner, "Mathematical Games," 120.

14. Dennett, *Darwin's Dangerous Idea*, 169.

15. Gardner, "Mathematical Games."

16. William Poundstone, *The Recursive Universe: Cosmic Complexity and the Limits of Scientific Knowledge* (New York: Morrow, 1985), 40.

17. Poundstone, *Recursive Universe*, 38, quoted in Dennett, *Darwin's Dangerous Idea*, 170–71.

18. Dennett, *Darwin's Dangerous Idea*, 171.

19. Hilary Putnam, "Reductionism and the Nature of Psychology," *Cognition* 2 (1973): 131–46.

20. Edward O. Wilson, *On Human Nature* (Cambridge, Mass.: Harvard University Press, 1978), 12–13.

21. Lindley Darden and Nancy Maull, "Interfield Theories," *Philosophy of Science* 44 (1977): 53.

22. John Dewey, *Logic: The Theory of Inquiry*, vol. 12, *John Dewey, The Later Works, 1925–1953*, ed. Jo Ann Boydston (Carbondale: Southern Illinois University Press, 1938), 26.

23. Dewey, *Logic*, 26.

24. Readers wishing to get their feet wet are referred to Nicholas Humphrey, *Consciousness Regained* (Oxford: Oxford University Press, 1984); Nicholas Humphrey, *A History of the Mind: Evolution and the Birth of Consciousness* (London: Chatto & Windus, 1992); Crick, *Astonishing Hypothesis*, 1994; Daniel Dennett, *Consciousness Explained* (Boston: Little Brown, 1991); Dennet, *Freedom Evolves*, and *Darwin's Dangerous Idea*; John Searle, *The Mystery of Consciousness* (New York: New York Review of Books 1997); John Searle, *Mind: A Brief Introduction* (New York: Oxford University Press, 2004); David Chalmers, *The Conscious Mind: In Search of a Fundamental Theory* (New York: Oxford University Press, 1996); V. S. Ramachandran and Sandra Blakeslee, *Phantoms in the Brain* (New York: Quill, 1988); Owen Flanagan, *Consciousness Reconsidered* (Cambridge, Mass.: MIT Press, 1992); Owen Flanagan, *The Problem of the Soul: Two Visions of Mind and How to Reconcile Them* (New York: Basic Books, 2002); Daniel Wegner, *The Illusion of Conscious Will* (Cambridge, Mass.: MIT Press, 2002); V. S. Ramachandran, *A Brief Tour of Human Consciousness: From Imposter Poodles to Purple Numbers* (New York: PI Press, 2004); and Christof Koch, *The Quest for Consciousness: A Neurobiological Approach* (Englewood, Colo.: Roberts & Company, 2004).

25. Originally presented in Daniel Dennett, *The Intentional Stance* (Cambridge, Mass.: MIT Press, 1987), 295–98; a revised form is found in Dennett, *Darwin's Dangerous Idea*, 422–27. Cf. Richard Dawkins, *The Selfish Gene* (New York: Oxford University Press, 1976), discussion of the problems faced by genes in designing mobile survival machines.

26. Dennett, *Darwin's Dangerous Idea*, 424–25.

27. Stephen Mithen, *The Prehistory of the Mind: The Cognitive Origins of Art and Science* (London: Thames & Hudson, 1996); cf. Edward Slingerland, *What Science Offers the Humanities: Integrating Body and Culture* (New York: Cambridge University Press, 2007), chap. 4, for a discussion of how the field of cognitive linguistics (conceptual metaphor and blending theory) provides helpful methodologies for tracing out the specifics of cross-domain mappings and the structure of blended domains.

28. Dennett, *Freedom Evolves*.

29. Dawkins, *Selfish Gene*.

30. Dennett, *Intentional Stance*.

31. The best general introduction to theory of mind, and the experimental evidence supporting its existence, is Paul Bloom, *Descartes' Baby: How the Science of Child Development Explains What Makes Us Human* (New York: Basic Books, 2004). Cf. Simon Baron-Cohen, *Mindblindness: An Essay on Autism and Theory of Mind* (Cambridge, Mass.: MIT Press, 1995); and the special edition of the *Journal of Cognition & Culture* 6 (2006): 1–2.

32. Michael Arbib, *In Search of the Person: Philosophical Explorations in Cognitive Science* (Amherst: University of Massachusetts Press, 1985), 115. Cf. Francis Crick's

comment that much of the brain's behavior is emergent, but not in any mystical sense: it can, at least in principle, be predicted from the nature of the parts involved combined with an account of how these parts interact. Crick, *Astonishing Hypothesis*, 11.

33. See, for instance, David Chalmers's comment that conscious experience is "a fundamental feature of the world, alongside mass, charge, and space-time." Chalmers, *Conscious Mind.*

34. Dennett's original position is laid out in Daniel Dennett, "Quining Qualia," in *Consciousness in Contemporary Science*, eds. Anthony Marcel and Edoardo Bisiach (New York: Oxford University Press, 1988); also see his debunking of Frank Jackson's (1982) qualia-based "Mary the color scientist" thought experiment (*Darwin's Dangerous Idea*, chap. 5). For a full account of Putnam's criticism of the general qualia argument, see Putnam, *The Threefold Cord: Mind, Body, and World* (New York: Columbia University Press, 1999), esp. 151–75.

35. Dennett, *Darwin's Dangerous Idea*, 111.

36. See Bloom, *Descartes' Baby*, xii–xiii.

37. See the discussion in Dennett, *Consciousness Explained*, chaps. 10–12; and Putnam, *The Threefold Cord*, 73–91.

38. Putnam, *The Threefold Cord*, 98.

39. Thomas Nagel, "What Is It Like to Be a Bat?" *Philosophical Review* 83 (1974): 448.

40. Timothy Wilson, *Strangers to Ourselves: Discovering the Adaptive Unconscious* (Cambridge, Mass.: Harvard University Press, 2002), provides a helpful survey of the relevant literature; cf. Jonathan Schooler and Charles Schreiber, "Experience, Meta-consciousness, and the Paradox of Introspection," *Journal of Consciousness Studies* 11 (2004): 17.

41. Schooler and Schreiber, "Experience," 17.

42. Searle, *Mind.* Daniel Dennett has been the most dogged and vociferous critic of Searle's position; see especially Dennett's *Consciousness Explained* and *Darwin's Dangerous Idea* (397–400).

43. Searle, *Mind*, 113.

44. Searle, *Mind*, 113.

45. Searle, *Mind*, 131.

46. Searle, *Mind*, 122.

47. Searle, *Mind*, 38.

48. Searle, *Mind*, 39.

49. John Searle, "Minds, Brains, and Programs," *Behavioral and Brain Sciences* 3 (1980): 417–57.

50. Arbib, *In Search*, 30.

51. Patricia Churchland, *Neurophilosophy: Toward a Unified Science of the Mind-Brain* (Cambridge, Mass.: Bradford Books / MIT Press, 1986), 315.

52. John Locke, *An Essay Concerning Human Understanding* (1690; Oxford: Clarendon Press, 1975).

53. Dennett, *Darwin's Dangerous Idea*, 3.

54. Dennett, *Darwin's Dangerous Idea*, 7.

55. Putnam, *The Threefold Cord*, 148.

56. Flanagan, *Problem of the Soul*, 65.

57. I here take issue with Searle's claim that physicalism functions as a modern religious dogma, accepted "without question" and with "quasireligious faith" (Searle, *Mind*, 48). No doubt some physicalists are dogmatists as well, but dogmatism is not intrinsic to the position. Searle's assertion that physicalism leaves out "some *essential* mental feature of the universe, which *we know*, independently of our philosophical commitments, *to exist*"—that it denies "the *obvious fact* that we all *intrinsically have* conscious states and intentional states" (49, emphases added)—seems to me much more faith-like than the claim defended by the likes of Dennett that physicalism just seems to be the best explanation that we have right now.

58. For instance, many recent attempts to escape the conclusions of physicalism focus on the phenomenon of indeterminacy at the quantum level, which seems to break us out of a deterministic universe. Certain prominent scientists, such as Roger Penrose (*The Emperor's New Mind: Concerning Computers, Minds, and the Laws of Physics* [New York: Oxford University Press, 1989]), argue that quantum indeterminacy is the locus of human free will. This seems, however, to be a fundamentally flawed and desperate strategy. To begin with, it is based on a distant and thin similarity between what we think human free will must be like and a completely unrelated phenomenon that shares one of the desired characteristics: indeterminancy. There is also the inconvenient fact that the wondrous quality of indeterminacy is present only at the quantum level: once we get up into levels that are humanly relevant, such as that of neurons or hormones, LaPlacian determinism reexerts its iron grip. The most basic and fatal flaw in this argument, however, is that indeterminacy is nothing more than randomness, which is not really what defenders of strong free will are after (Searle, *Mind*, 24–25). The human conception of free will requires that this will be determined by *something*—reasons, desires, spontaneous impulses, etc. Free will as utter randomness is as horrific a concept at a human level as the deterministic absence of free will. For cogent discussions of Penrose's position, see Dennett, *Darwin's Dangerous Idea*, 428–51; and Flanagan, *Problem of the Soul*, 127–32.

59. Flanagan, *Problem of the Soul*, xiii.

60. Flanagan, *Problem of the Soul*, 20.

61. Flanagan, *Problem of the Soul*, 9–10.

62. Dennett, *Darwin's Dangerous Idea*, 19. Cf. Dennett's comment regarding the "Zombie hunch" (the feeling that its makes sense to speak of consciousness as something over and above the functioning of the embodied mind): "If you are patient and open minded, it will pass" (23).

63. Dennett, *Darwin's Dangerous Idea*, 17.

64. Flanagan, *Problem of the Soul*, 8.

65. Archaic modern humans have been burying their dead for at least ninety-two thousand years (Ofer Bar-Yosef, *Human Migrations in Prehistory: The Cultural Record*, a presentation at World Knowledge Dialogue, Crans-Montana, Switzerland, September 15, 2006). Elaborate, ritualized burial is as good a litmus test as any of the presence of theory of mind. When an implement breaks, you throw it away, and the remains of living prey are disposed of as quickly and conveniently as possible. Special treatment of the human corpse indicates that a shift has occurred and that

the human body is now being viewed as linked to something fundamentally distinct from objects.

66. Bloom, *Descartes' Baby.*

67. Mencius, *Mencius*, trans. D. C. Lau (London: Penguin, 1970), 168.

68. Stewart Guthrie, *Faces in the Clouds: A New Theory of Religion* (Oxford: Oxford University Press, 1993), 7.

69. Guthrie, *Faces*, 5.

70. Deborah Kelemen: "Function, Goals, and Intention: Children's Teleological Reasoning about Objects," *Trends in Cognitive Sciences* 3 (1999a): 461–68; "Why Are Rocks Pointy? Children's Preference for Teleological Explanations of the Natural World," *Developmental Psychology* 35 (1999): 1440–452; "British and American Children's Preferences for Teleo-functional Explanations," *Cognition* 88 (2003): 201–21; "Are Children 'Intuitive Theists'? Reasoning and Purpose and Design in Nature," *Psychological Science* 15 (2004): 295–301.

71. Cf. Jesse Bering's work on the attribution of supernatural agency by adults and children to explain otherwise mysterious, causeless events. Jesse Bering, "The Existential Theory of Mind," *Review of General Psychology* 6 (2002): 3–24; Jesse Bering and Becky Parker, "Children's Attributions of Intentions to an Invisible Agent," *Developmental Psychology* 42 (2006): 253–62.

72. Bering, "Existential Theory," 126.

73. See, for instance, Francisco Varela, Evan Thompson, and Eleanor Rosch, *The Embodied Mind: Cognitive Science and Human Experience* (Cambridge, Mass.: MIT Press, 1991).

74. For a representative sampling, see Ainslie Embry, ed., *Sources of Indian Tradition*, 2nd ed., vol. 1 (New York: Columbia University Press, 1998), 103–108.

75. See D. Jason Slone, *Theological Incorrectness: Why Religious People Believe What They Shouldn't* (New York: Oxford University Press, 2004), for a characterization of such tensions between doctrine and intuitive concepts as instances of "theological incorrectness."

76. From the album *I Put a Spell on You* (1965), song written by Anthony Newley and Leslie Bricusse for the 1966 musical *The Roar of the Greasepaint—the Smell of the Crowd.*

77. A chilling literary portrait of such a state is provided in William Styron, *Darkness Visible* (New York: Random House, 1992).

78. Charles Taylor, *Sources of the Self: The Makings of Modern Identity* (Cambridge, Mass.: Harvard University Press, 1989), 57.

79. Taylor, *Sources*, 59.

80. Wegner, *Illusion.*

81. Cf. Dennett's characterization of free will as "an evolved creation of human activity and beliefs" and therefore "just as real as such other human creations as music and money" (*Freedom Evolves*, 13). Also see John Searle's discussion of the inescapability of humanly created reality, the features of which appear "as natural to us as stones and water and trees" (*Mind*, 4).

82. R. James Blair, "Neurocognitive Models of Aggression, the Antisocial Personality Disorders, and Psychopathy," *Journal of Neurology, Neurosurgery, and Psychiatry* 71 (2001): 727–31.

83. Pinker, *Blank Slate*, 193. Cf. Michael Ruse's comment that, "as a function of our biology, our moral ideas are thrust upon us, rather than being things needing or allowing decision at an individual level" (Michael Ruse, *Taking Darwin Seriously: A Naturalistic Approach to Philosophy*, 2nd ed., [Oxford: Blackwell, 1998], 259).

84. Pinker, *Blank Slate*, 32.

85. Pinker, *Blank Slate*, 26.

86. This is essentially the same insight behind Searle's granting of special ontological status to human consciousness.

87. For a recent study illustrating this phenomenon, see Jesse Preston and Nicholas Epley, "Explanations versus Applications: The Explanatory Power of Valuable Beliefs," *Psychological Science* 16 (2005): 826–32.

88. The American Association for the Advancement of Science.

89. *Exapted* means to use a feature that once served adaptation for new purposes. See Stephen Jay Gould and Elizabeth S. Vrba, "Exaptation: A Missing Term in the Science of Form," *Paleobiology* 8, no. 1 (1982): 4–15.

90. Crick, *Astonishing Hypothesis*, 258.

91. John 17:14–15.

92. Burton Watson, trans., *Zhuangzi: Basic Writings* (New York: Columbia University Press, 2003), 40–41.

93. Immanuel Kant, *Groundwork of the Metaphysic of Morals*, trans. H. J. Paton (New York: Harper Torchbooks, 1964), 115–16.

94. Of course, what counts as human is up for grabs, and the idea that every member of the biological species *Homo sapiens* is human is a relatively recent idea—the category has historically tended to encompass only one's own tribe. The recurrent reality of genocide, even in our modern world, serves as a chilling reminder of how quickly and easily groups formerly seen as humans can be reclassified as things.

95. Ian McEwan, *Saturday* (Toronto: Alfred A Knopf, 2005), 56. Used by permission.

5

Downward Causation and Religion

Alicia Juarrero

E VER SINCE the mid-seventeenth-century's scientific revolution, wholes have been considered epiphonemena with no causal efficacy. Modern science and philosophy's unstated presupposition that causal power operates only as efficient causality turned meaning and meaningful behavior into problems in need of explanation. How could the forceful interaction of passive and mindless material particles produce thought and feeling? The only answer possible was either in terms of a thoroughgoing materialism or some form of dualism that postulated that cognition and action (not mere behavior) must be the effects of a substance that, because it belonged to a different ontological order, could cause such effects.

It wasn't a clean break with tradition, however. Modern philosophy and science inherited a peculiar mélange of ideas concerning causality from the combined traditions of Aristotle and Newton: from the tradition of mechanism the principle that only efficient causes operating as push-pull forces are causes at all.

For Aristotle, because all change is a transformation from potentiality into actuality, accounting for any particular instance of change requires clear specification of the role each of the four causes (final, formal, efficient, and material) plays in making the potential actual. However, since in no instance of motion does the principle of movement lie "in the thing itself qua thing, *nothing can cause itself.*"[1]

As is well known, however, modern science discarded the concept of formal cause by insisting that organic wholes, no less than aggregates, reduce to their component parts. Newtonian mechanics also left no room for end states to serve as intentional objects of desire or goals of action. Aristotelian final causality was, accordingly, also discarded. As a result, by ignoring both Newton's belief that gravity is a form of action-at-a-distance and his recognition of the three-body problem, the theoretical framework that came to be known as Newtonianism or Mechanism becomes identified almost by definition with the view that physical processes occur deterministically by the direct transmission of force among discrete bodies in contact. The example of billiard balls bumping into each other is often used to illustrate this form of cause, which since the seventeenth century has come to be identified with causality *tout court.*

In a scientific worldview that considers only point masses and forces to be real, qualities such as temperature and color end up being dismissed as subjective. By the end of the seventeenth century, all relational properties are relegated to the inferior status of "secondary qualities." In consequence, context was left with no role to play; indexicals lose their claim on reality as situatedness and point of view become secondary as well. The arrow of time embodied disappears from time-reversible equations, leaving the reaching out and aimed-for goal-directness of purpose and teleology a puzzle that, like the ether, call out either for some sort of dualism or for a materialist reduction to mechanistic forces on point masses.

It is easy today, with the clarity of hindsight, to find fault with this peculiar combination of the Aristotelian and Newtonian principles that nothing causes itself and that all causality is efficient causality. But the combination worked so well for teasing out the laws of planetary motion that it was said that Newton had read the mind of God.

As always in the history of science, anomalies began to mar the picture of a tidy clockwork universe.

The Aristotelian-Newtonian combo on causality first proved to be recalcitrant when applied to biology. The causality at work in this domain did not fit into the mold of efficient causality; moreover, living things evidently caused themselves, and not qua other, as Aristotle required. The problem of what has been called *interlevel* causation—how elements interact to produce systemic wholes, which in turn affect the behavior of the components that make them up—has been a puzzle to thinkers ever since.

In the last half-century, however, nonlinear, far-from-equilibrium thermodynamics and complex adaptive systems theory have revitalized this topic through a deeper appreciation of the way autocatalytic cycles and self-organizing processes operate. A better understanding of how boundary conditions can influence components of systemic wholes (as opposed to aggregates) has done the same for the concept of downward causation by rethinking the old discarded notions of final and formal cause. Because complex adaptive systems display emergent characteristics that are irreducible to those of the sum of their components, as a by-product of this new conceptualization of top-down causality, the notions of mind, consciousness, and cognition have been reintroduced into respectable company after centuries of being actively shunned by philosophers and scientists alike.[2]

In this chapter I explore the concept of downward causation and discuss how the approach employed by the new science of complex adaptive systems toward mereological causation has revitalized cognitive studies. I also critically review philosophers such as Immanuel Kant, John Stuart Mill, and Charles Sanders Peirce, who questioned the universality of mechanistic causality, and early twentieth-century British emergentists such as Samuel Alexander and C. Lloyd Morgan, who argued that taking top-down causation at the very least makes room for some sort of deity. Changing our understanding of the causal role the environment plays in biological development and evolution—not as efficient cause, however—has been a key turning point in cognitive studies.

Immanuel Kant

Objections to restricting all causation to a mechanistic push-pull efficient causality appeared soon after Newton's formulation of mechanism. Despite claiming to have been awakened from his dogmatic slumbers by Hume, Immanuel Kant's category of causality owes more to Leibniz than to Hume. Before embarking on his critical turn, Kant (1724–1804) had defended a Leibnizian notion of *vis viva*. Mechanism views forces as affecting a passive body from the outside, but if substances cannot *act* "there would be no extension and, consequently, no space."[3] Despite repudiating this approach in his critical works later on, when Kant considers the self-organization present in living things he turns reluctantly to Leibniz.[4] Leibniz, who himself had maintained in an early paper that only mechanical explanation can supply us with a *vera causa*,[5] subsequently changed his mind and noted that, although mechanical principles may be sufficient to explain proximate efficient causes, "the origin of this mechanism itself has come not from a material principle and mathematical reasons alone but from a higher and, so to speak, a metaphysical source";[6] final cause must be invoked at this point, Leibniz had maintained.

Kant recognized that, contrary to the principles of Newtonian mechanics, organisms exhibit a built-in force. Things or objects, according to Kant, display two kinds of finality, relative and intrinsic. Whereas the former describes an object's adaptability as a means to other ends, *intrinsic* finality describes the relationship between a cause and its effects. As examples of the former Kant cites sandy soil, which promotes the growth of pine trees, and rivers, whose course and alluvial deposits benefit plant growth, which in turn benefits human beings. But both of these are quite unlike trees, which are clearly "both causes and effects of themselves."[7] A tree's leaves are not only produced by but also maintain the tree. The growth and maturation of living things, Kant saw clearly, are evidence of a type of recursive causality whereby "the genus, now as effect, now as cause, continually generated from itself and likewise generating itself, preserves itself generically."[8] And it is this type of causality—wherein a physical end is both cause and effect of itself—that is properly termed *final*, in Kant's view. He points out that the parts of living things, both as to their existence and form, are "only possible by their relation to the whole."[9] It is in this sense that only organisms exhibit *intrinsic finality*, which they do by virtue of their self-organizing capacity.

It is precisely the peculiar, recursive causality of self-organization that makes Kant question Newtonian mechanics, especially the second law of mechanics, which, following Aristotle's principle that nothing causes itself, holds that a body tends to remain at rest or in motion unless acted upon by an *external* force. The causality discoverable by mechanistic causality pertains to objects and things that are "only possible by means of something else as their cause."[10] The unity of principle found in organisms, on the other hand, "involves regressive as well as progressive dependency,"[11] a form of causality, Kant puzzles, that is "nothing analogous to any causality known to us."[12] *Organization* and its cognates, such as *organism*, Kant noted, refer to a structure wherein a member is not only a means but also an end; it both contributes to the whole and is defined by it.[13] No machine exhibits this kind of organization. Consistent self-organization constitutes a principle of unity that could not come about by mechanical processes, which are contingent in the sense of effecting causal change by means of outside forces. Accordingly, characteristics preserved by heredity and evolution cannot be accounted for mechanistically; they must be understood in terms of final causality, Kant insisted.

What the *Third Critique* offers twenty-first-century readers is a modern philosopher's acknowledgement of the top-down causal role of organization. But it is not only top-down; Kant understood that the higher organismic or systemic level emerges from—because it is composed of—interactions among parts. As is well known, however, Kant's critical program sees no other way out than to relegate finality to the regulative judgment; intrinsic teleology is therefore described as something we just have no choice but to *ascribe* to certain processes. Ascriptions of finality for Kant, as ascriptions of intentionality for Daniel Dennett, then, are merely epistemological stances, not constitutive or ontological claims. Absent any breakthrough in our scientific understanding of mereological causality—interlevel parts-to-whole or whole-to-parts causal connections—matters would remain by and large unchanged for over 150 years, with only a few but notable exceptions.

John Stuart Mill

Born only two years after Kant's death, John Stuart Mill (1806–1873) returned to the problem that bedeviled the Koenisburg philosopher—the relationship between Newtonian mechanics and causality. In particular, Mill objected to what he called the Principle of the Composition of Forces or Causes, and to the Doctrine of Proportionality, both of which he considered basic axioms of mechanism. According to the former, laws that apply in situations where several agents (or causes) contribute to the production of an effect merely "sum up" the combined influence of several causes acting separately. Allegedly, the principle of Composition of Forces not only describes mechanical phenomena, but since the effects of the combinations can be deduced from laws governing the causal agents separately, the principle also presumably accounts for mechanics as a demonstrative science. The Principle of Proportionality, which maintains that effects are proportional to their causes, is a related axiom that presupposes that the world is linear.

Whereas Kant chose examples from biology that did not fit mechanistic principles, Mill looked to chemistry for counterexamples to these two principles. Water, the effect of the chemical combination of hydrogen and oxygen, Mill noted, has properties different from the mere sum of these two components. The wetness of a molecule of water is a qualitative characteristic that cannot be deduced from the properties of its atomic components. The same is true for the sugary taste of lead and the blue color of vitriol.

With respect to the Principle of Proportionality, Mill pointed out that this doctrine does not apply to any phenomena—such as chemical phenomena—where increasing the cause alters the *kind* of effect (that is, in those cases where "a totally different set of phenomena arise.")[14] Not only are these new properties and entities not merely quantitatively different from what went before, they also bring qualitatively different laws into being. These examples from chemistry, Mill argued, clearly refute the universality of the foundations of mechanism. Something else is at work in nature.

It appeared as if the topic of emergent properties with top-down causality came up every quarter century or so.

Charles Sanders Peirce

Borrowing from Mill's objections to claims concerning the linearity of nature, Charles S. Peirce (1839–1914) renewed the argument by noting that the increasing complexity and order evident throughout nature and cosmological evolution belie the received axiom that "nature is uniform."[15] If *every event has a cause* implies that all causality operates in the deterministic manner of mechanism, then when this principle is combined with the second law of thermodynamics, how is nature's creativity to be explained? Some mechanism must be present to counteract the second law of thermodynamics' dissipation of energy and its relentless march toward heat death of the universe. The answer, Peirce suggests, must be "that *chance*, in the Aristotelian sense, mere absence of cause, has to be admitted as having some slight place in the universe"; perhaps randomness may be the "one essential agency on which the whole process depends."[16]

Comparing his views to those of Darwin and therefore suggesting a mechanism for the evolution of emergent properties, Peirce noted that the second law and Boyle and Charles's laws of gasses are "the results of chance—statistical facts, so to say. . . . I cannot help believing that more of the molecular laws . . . will be found to involve the same element, especially as almost all these laws present the peculiarity of not being rigidly exact."[17] The strict determinism of mechanical laws was being called into question.

But chance alone can't account for the appearance of the laws of physics; Peirce needed another principle for this to occur. He found the answer in the workings of what he called habit. If one postulates, he argued, that "the main element of habit is the tendency to repeat any action that has been performed before," then "winning" or "good" chance outcomes will be reinforced, and "losing" or "bad" chance outcomes would be weeded out—by habit. "Chance in its action tends to destroy the weak and increase the average strength of the objects remaining. Systems or compounds that have bad habits are quickly destroyed; those that have no habits follow the same course; only those that have good habits tend to survive."[18] "I will suppose that all known laws are due to chance and repose upon others far less rigid themselves due to chance and so on in an infinite regress, the further we go back the more indefinite being the nature of the laws." The laws of physics we see today, Peirce concluded, might simply "be habits gradually acquired by systems."[19] Substitute the word *selection* for *habit*, and Darwin's theory of variation and selection is not far away—not to mention Donald Hebb's principle that "neurons that fire together wire together."[20]

In his most recent book, *Cosmic Jackpot*, Paul Davies echoes Peirce when he suggests that "abandoning Platonism would make room for teleology."[21] Davies quotes Rolf Landauer,[22] among others, as questioning the universality and immutability of the laws of physics, suggesting instead that perhaps these "come with an inbuilt level of looseness or flexibility—a level that is minuscule today but significantly higher in the universe's very early moments."[23] Allowing a certain amount of wiggle room in the laws of physics, Davis maintains, would provide them with a considerable "degree of looseness or ambiguity," enough to make room for a teleological biofriendliness toward the evolution of life and mind.[24] Make room for final and formal causality, that is.

But questions remain with this approach. Both Landauer and Peirce postulate that it is in the earliest moments of the universe that the laws of physics were least rigid; over time, Davis speculates, natural laws—or habits, to use Peirce's term—became increasingly exact until they reached the strict determinism of mechanism. However, the trend seems to operate in reverse once evolution produced living things. Within biology, moreover, the behavior of organisms belonging to the more recently emerged taxa is less-rigidly programmed than that of organisms belonging to earlier taxa. And it is in the so-called human special sciences of sociology, anthropology, economics, and history that the wiggle room is greatest. So which came first, the rigidity or the wiggle room?

But back to the history lesson.

The British Emergentists

C. Lloyd Morgan (1852–1936), one of the so-called British Emergentists of the early twentieth century, was the central figure of a group of thinkers that included C. S. Lewis and Samuel Alexander, all of whom maintained that emergent evolution is in play whenever new and unpredictable *types* of existents emerge. Unlike continental thinkers like Henri Bergson, who accounted for the process by appealing to the deus ex machina operation of some *élan vital*, Morgan recognized that, for the most part, the distinctive quality that appears in phenomena we call emergent is in fact a new kind of *relation*; the emergent is the embodiment of "an integration achieved by a supervenient level of relatedness."[25] In this Morgan follows Alexander before him, who insisted that each higher entity is an emergent "complexus" of lower entities.

Alexander had called the universe's inward drive toward increasingly higher entities *nisus toward deity*. Anticipating twentieth-century philosophers such as Jaegwon Kim and Donald Davidson's writings on supervenience, Alexander maintained that the higher-level complexus "involves" the lower but is a distinct level of its own. For Morgan, however, claiming that the novel properties that emerge from integration "can only be explained by invoking some chemical force, some *élan vital*, some entelechy in some sense extranatural," amounts to "questionable metaphysics."[26] So if it is neither some extranatural force nor some external mechanism that explains the appearance of novel properties, what, then, is the causal agent involved, and what kind of causality is at work here? The possibility of top-down causality surfaces once again.

Despite agreeing with Lewes and Alexander on the reality of emergence, while at the same time taking care not to fall prey to questionable metaphysics, Morgan argues that there is nothing in emergent evolution that *precludes* an appeal to God. "Unlike Alexander, who works bottom-up and insists that the quality at each higher level is emergent from the lower, Morgan bases his argument on an examination of what we today would call top-down causality: the descending order of the pyramid, so to speak, from higher to lower."[27]

If the emergent higher level is to have causal efficacy, the relatedness that constitutes the higher level must alter the "existing go of events" at the component level.

It is the "new manner in which lower events happen" that accounts for novelty in evolution. Morgan offers the example of perception, which depends on the guidance of consciousness. New relations on the higher level "guide and sustain the course of events distinctive of that level." Even though there would be no higher level without the "involvement" of the goings-on at the lower, it remains true that these lower-level events "run their course" differently than they would have had the higher supervenient level been absent. What kind of causality is operative here? Morgan does not say, but the way in which he describes the "guidance" and "sustenance" that the higher level provides to the lower clearly has more in common with Aristotelian formal cause than with the efficient, forceful cause of mechanics. Finding a way to account for the apparent ubiquity of top-down causality remains a central topic in the philosophy of science.

Since Morgan claims that the lower level's "dependence" on the novel and emergent higher level is no less essential than the "involution" of the higher level on the lower, he also concludes that, although it is "beyond proof," postulating the existence of a God, here understood as the highest guiding, sustaining, and directive Activity on which all other levels below depend.[28]

It is clear that for Alexander, Deity appears only at the end of the process with the emergence of the highest relational level; Morgan's emphasis on the descending power of the pyramid suggests that he understands God as actively sustaining all instances of top-down causation. Whether Morgan thought top-down causality implied that such a God would necessarily or possibly preexist the cosmos, however, is unclear. The difference between Morgan and Alexander's positions is a subtle one, but Alexander's position appears to be more similar to that of thinkers like Stuart Kauffman—Alexander's "*nisus* of the universe pressing on to levels as yet unattained" is reminiscent of Kauffman's ubiquitous propensity or thrust toward "self-organization."[29]

In any case, Aristotle and Newton must be turning in their graves: their cherished axiom, nothing causes itself, is no longer the self-evident principle they held it to be.

Complex Dynamical Systems Theory

Morgan and Alexander often foreshadow many of the concepts employed in nonlinear dynamical systems theory. It wasn't until the second half of the twentieth century that topics such as self-organization in general, and the self-renewal and self-regulation of living things in particular, were directly addressed by scientists, after which books like Douglas Hofstadter's *Gödel, Escher, Bach* quickly took up the subject of circular causality. I date the shift to Ilya Prigogine's theory about dissipative structures in chemical-reaction systems: The possibility of nonlinear causality was broached once Prigogine's team empirically confirmed that the "dynamics of the [autopoietic whole] provides the framework for the behavioral characteristics and activities of the parts."[30] In light of that research it was impossible to deny that dissipative structures display a form of top-down causality. The whole has causal efficacy on the components at the lower levels; each of the processes constituting *auto-poiesis* "recursively generates and recovers the same complex of processes that produced them."[31] The unusual role of autocatalysis in many self-organizing structures such as the Belousov-Zhabotinsky reaction clearly satisfies Kant's definition of an intrinsic physical end: the sequence of reactions in the B-Z reaction both produces itself and maintains itself as itself despite a constant turnover of its components. What we have in *autopoiesis*, in my judgment, is as much a reformulation of Aristotelian formal causality as of teleology and final causality. *Autopoiesis* allowed scientists and philosophers to rethink formal causality in scientifically respectable terms.

Despite all that, however, philosophers and scientists across the board continued to resist labeling *causal* the influence that boundary conditions of systemic wholes effect on their components. Some of this resistance is related to other assumptions about causality such as that true causes are deterministic and strictly lawlike,[32] but I suspect that a lot of it is due to the fact that philosophers have uncritically accepted a mechanistic paradigm of causality. Instead, without defense, they resort to verbs such as *modulate, regulate*, and *constrain* wherever and whenever this kind of top-down causal influence is present. It is undeniable, however, that the top-down power present in dissipative structures does capture a central aspect of causality, what Lewis calls the sine qua non nature, or necessary condition role, of causes: if the self-organized boundary conditions of autopoietic dissipative structures did not actively influence as they do the components that make it up, the overall dynamic system would be other than it is.

So let us agree to use the term *constraint* whenever this kind of mereological or interlevel causality is involved. How would constraints work, especially in light of the fact that, as complexity increases over time, constraints do not just close off options but must open up alternatives as well?

The first point to make is that, contra Newtonian science and modern philosophy's refusal to countenance relational properties, constraints are features either of an object's connections with the environment or of its embeddedness via feedback in that environment. Constraints are clearly not

primary qualities, such as mass, that objects display regardless of circumstances. Instead, it is the orderly context in which components are embedded
that constrains them. Constraints are therefore relational properties that parts
acquire in virtue of being unified—not just aggregated—into a systematic
whole. There is no organism without organization, and there is no organization if particles or elements are independent and isolated from each other.
Interaction is required.

When previously independent and isolated particles become interconnected—truly interconnected, not just clumped together or pushed by the
other—the marginal probability of the behavior of those heretofore particles
changes: by definition things like catalysts and feedback loops increase the
probability that a chemical reaction will occur. By doing so, catalysts and
feedback loops truly link together phenomena into a systematic whole. When
previously unrelated things become correlated given the presence of catalysts
and/or feedback loops, qualitatively different things suddenly become possible. Catalysts and feedback loops are thus *context-sensitive constraints* that
take molecules away from independence; they are the mechanisms whereby
differentiation and complexification occur. *First-order contextual constraints*,
such as catalysts and feedback loop, are thus bottom-up enabling constraints—by precipitating self-organization they free up degrees of freedom
such that the newly emergent systemic whole can now do more than the sum
of its components (without being other than its components).

True Self-Cause: Bottom-Up Causality

I hazard a speculative aside here. The wave-particle-duality conundrum of
quantum mechanics—whether the phenomenon manifests itself as a particle
or a wave depending on the type of *experiment* or *observation* performed—has
occasioned tremendous debate on what counts as an *observation*. Most commentators do not require the presence of a conscious mind; in this context,
phrases like *interaction with a measuring apparatus* are synonymous with
observation. Whose observation? Does this phenomenon leave room for a
Berkeley-type deity? Toward the end of *Cosmic Jackpot* Davis cites Wheeler's
claim that although life and mind are undoubtedly products of the cosmos,
the observer problem in the two-slit experiment of quantum physics suggests
that the sequence *cosmos → life → mind* might in fact be a closed loop with
life and mind also being as indispensable to physics and the cosmos as these
two are to the appearance of life and mind.

Discussions of this Schroedinger's cat problem have for the most part
concentrated on the *measuring* or *observation* terms in the phrase *interaction*

with a measuring apparatus. Nonlinear dynamical systems theory might be pointing us to the other term, *interaction,* as perhaps the more fruitful avenue of exploration. As is the case with the quite different results that obtain in the *iterated* version of the Prisoner's Dilemma, the bottom-up causal role of feedback, connectivity and relatedness, not just energetic force, is undeniable. Whether or not speculative or transcendental religions are right in postulating an external teleological goal toward which nature aims, could nature's self-organizing propensity provide the ontological support for values, a support that Hans Jonas finds missing in mechanistic science?[33]

In any case, once the systematic whole of, for example, a B-Z waveform organizes as a result of first-order context-sensitive constraints, a phase change has taken place:[34] the autocatalytic network's organization as a whole suddenly emerges as a contextual constraint on its components. I call these *second-order contextual constraints.* These top-down contextual constraints serve as—indeed *are*—the boundary conditions in which the components are located and to which they are now systematically and not just externally related. Mechanical virtual governors are also examples of such top-down, second-order, contextual constraints. As such, the newly created systematic whole imposes—top-down—second-order constraints on its components, by which I mean that the components' degrees of freedom are now limited, given the systemic context in which they are embedded. (To avoid reification, it is perhaps more accurate to say that the whole *is* the second-order contextual constraints on the components.) Once captured in a dynamical system these no longer have the alternatives or the freedom they did when they were independent and uncorrelated. Top-down constraints are therefore *limiting* constraints.

True Self-Cause: Top-Down, from Whole to Parts

It is thanks to this renewed understanding of how mereological causation operates that the theories of self-organization and nonlinear dynamical-systems theory have provided philosophers and scientists with a scientifically respectable account of both circular causality and the intuition that wholes are more than the sum of their parts—without appealing to external final causes, or hypothetical *élans vitals.* The meaning of *more* implied by these theories is reminiscent of what philosophers call *property,* as opposed to substance dualism: *qualities* possessed by the whole are different from and not logically deducible from those of the parts.

The importance of interconnectivity for the emergence of wholes with novel properties can be illustrated with an analogy involving neural networks.

Artificial neural networks with connections from later units back to earlier units (such as occurs in the human neocortex) are called *recurrent*. In contrast to networks without such recurrence, the output of recurrent networks is dynamic—they create what complexity and chaos theorists call *attractors*: the connection weights of recurrent networks recalibrate and thereby self-organize dynamic subregions within the system's state space.

Hinton, Plaut, and Shallice designed a recurrent neural network to read words.[35] When the researchers then lesioned the network upstream from the feedback units, the system's output often included visual errors rather like those of patients with surface dyslexia: when presented with the word *cot*, the network's output might be *cat*. But lesions either downstream from the feedback units or to the feedback units themselves produced astonishing errors: when presented with the word *cot*, the network might print out *bed*! To explain this remarkable result, which parallels that of patients with so-called deep dyslexia, Hinton et al. postulate that, thanks to the feedback's connectivity, the network's hidden units must have self-organized a semantic attractor—that is, a self-organized space with emergent properties that can only be characterized as semantic because they embody the word's meaning or sense in the organization of the relationships that constitute the higher-dimensional attractor.

From the perspective of the framework presented in this paper, Hinton, Plaut, and Shallice's recurrent artificial network's output originated in a self-organized dynamical space whose second-order, context-dependent constraints, *as embodying semantics*, guide the network's output within that space. When the network reads the word correctly, the output represents the endpoint of a trajectory through this high-dimensional, semantically organized space. And it is in virtue of their high-level properties—that is, of the *meaning* embodied in the relations among the components—that the output is produced. In short, the output is *constrained by semantics*, because the organization's dynamics are causal as *meaningful*. Because the output originates in a higher-dimensional level of organization, we have here an example of the top-down causality of second-order, context-sensitive constraints.

In turn, what dynamical systems theory implies for the philosophy of mind is that once a semantic state space self-organizes in the human brain, trajectories within that higher dimensional, neurological space embody new rules: they embody the constraints of logic, meaning, value, and the like. The trajectory is now constrained semantically. The difference between the variation in neuronal activity before and after semantic attractors self-organize measures the degree to which those rules constrain the system's behavior. Thinking of top-down causality in this manner makes room for a reconceptualization of morality and free will: a free person is, on this account, the person whose behavior is constrained by those higher levels of meaning and value, not the person whose behavior is uncaused. "Freedom is choosing meaningfully, not spontaneously."[36]

Was C. Lloyd Morgan right? Does the reality top-down causation open up the possibility of an appeal to an ultimate top-down agency? There appears to be no role in self-organized dynamical systems theory for a top-down agency that preexists the universe. The top-down causal influence of wholes on their parts is nothing *other than* the second-order context-sensitive constraints embodied in the interdependence among the components. The newfound scientific respectability of top-down causality in the sense of the operation of these constraints is not equivalent to, and does not require for explanatory consistency, a transcendent deity as plays a part in what Philip Kitcher calls *supernaturalist* or *providentialist* religions. Given the function of randomness and probability in the stochastic version of complexity theory, neither does this science suggest a particular direction or goal to the evolution of the universe.

But given that the trajectory of cosmology and evolution in fact led to the appearance of such phenomena as values, ethics, and morals, the ontological reality of top-down causality operating at each step of the hierarchy does appear to leave room for what Philip Kitcher calls *spiritual religion*, a set of beliefs that leave aside claims pertaining to the afterlife and a transcendent Creator and place in their stead a faith in "ethical models of right action and moving portraits of nobly lived lives."[37] And grounding such values on the ubiquitous connectedness and interdependence that made the cosmos possible would broaden the traditional concept of ethics to include norms pertaining to ecological issues. Now that there is no denying that top-down causality, in the sense described above, is ontologically real, there is even less reason to deny the causal efficacy of such ideals, which after all can now be understood to embody the "integration achieved by a supervenient level of relatedness," in Morgan's words.

Along with Kitcher, however, I leave it to the defenders of spiritual religion to determine if such a doctrine really amounts to a religion proper or to merely "secular humanism viewed through stained glass."[38]

Notes

1. Aristotle, *Metaphysics*, book 9.8, 1049b6–11.

2. By *irreducible complexity* I mean only that higher-level properties cannot be deduced logically from lower level laws.

3. Immanuel Kant, "Thoughts on the True Estimation of Living Forces, and Criticism of the Proofs Propounded by Herr von Leibniz and Other Mechanists in Their

Treatment of This Controversial Subject, Together with Some Remarks Bearing upon Force in Bodies in General," in *Inaugural Dissertation and Early Writings on Space*, trans. John Handyside (Chicago: Open Court, 1929), 870.

4. See Alicia Juarrero-Roqué, "Teleology and Modern Chemistry," *The Review of Metaphysics* 39, no. 1 (1985): 107–35.

5. Gottfried Wilhelm Lebiniz, *Philosophical Papers and Letters*, vol. 2, trans. Leroy Loemker (Chicago: University of Chicago Press, 1956), 266–67.

6. Leibniz, *Philosophical Papers*, 810–11.

7. Kant, "Critique," 18 (§64, Ak. V, 371). All subsequent references to Kant are from "The Critique of Teleological Judgement," in *The Critique of Judgement*, trans. J. C. Meredith (Oxford: Clarendon Press, 1980). References after "Ak" cite Kant's *Gesammelte Schriften*, prepared under the supervision of the Berlin Academy of Sciences.

8. Kant, "Critique," 18 (§64, Ak. V, 371).

9. Kant, "Critique," 20 (§65, Ak. V, 373).

10. Kant, "Critique," 116 (§87, Ak. V, 448).

11. Kant, "Critique," 25–26 (§5, Ak. V, 337).

12. Kant, "Critique," 23 (§65, Ak. V, 375).

13. Kant, "Critique," 23 (§65, Ak. V, 375).

14. John Stuart Mill, "On the Composition of Causes," reprinted in A. Juarrero and C. Rubino, eds., *Emergence, Self-Organization, and Complexity: The Anticipatory Works* (Litchfield, Ariz.: Isce Publishing, 2008).

15. Charles S. Peirce, "Design and Chance," in *Writings of Charles S. Peirce: A Chronological Edition*, vol. 4, ed. Christian J. W. Kloesel et al. (Bloomington: Indiana University Press, 1986), 544–54. Reprinted in Juarrero and Rubino, *Emergence*, 49.

16. Peirce, "Design and Chance," 549.

17. Peirce, "Design and Chance," 544–54.

18. Peirce, "Design and Chance," 553.

19. Peirce, "Design and Chance," 54.

20. Donald Hebb, *The Organization of Behavior* (New York: Wiley, 1949), vii.

21. Paul Davies, *Cosmic Jackpot* (New York: Houghton Mifflin, 2007), 127.

22. Rolf Landauer, "Wanted: A Physically Possible Theory of Physics," *IEEE Spectrum* 4, no. 9 (1967): 105.

23. Davies, *Cosmic Jackpot*, 248.

24. Davies, *Cosmic Jackpot*, 127.

25. C. Lloyd Morgan in Juarrero and Rubino, *Emergence*, 15.

26. Reprinted in Juarrero and Rubino, *Emergence*. Whether Morgan's interpretation of Alexander is correct, that *nisus toward deity* constitutes an extranatural entelechy, is debatable.

27. Juarrero and Rubino, introduction, *Emergence*.

28. Juarrero and Rubino, *Emergence*.

29. Juarrero and Rubino, *Emergence*.

30. Milan Zeleny, "Autopoeisis, a Paradigm Lost?" in *Autopoeisis, Dissipative Structures, and Spontaneous Social Orders*, ed. Milan Zeleny (Boulder, Colo.: Westview Press, 1980), 27.

31. Milan Zeleny, "Autopoiesis, a Paradigm Lost?" in *Autopoiesis, Dissipative Structures, and Spontaneous Social Orders*, ed. Milan Zeleny (Boulder, Colo.: Westview Press, 1980), 27.

32. Donald Davidson, "Actions, Reasons and Causes," in *Essays on Action and Events*, ed. Donald Davidson (Oxford: Oxford University Press, 1980).

33. H. Jonas, "Gnosticism, Existenalism, and Nihilism," in *The Phenomenon of Life* (Evanston, Ill.: Northwestern University Press, 1966).

34. It is perhaps more accurate to say that the closure of the first-order context-sensitive constraints *is* the phase change.

35. G. E. D. Hinton, C. Plaut, and T. Shallice, "Simulating Brain Damage," *Scientific American* 269 (1993): 76–82.

36. Robert Artigiani, "Societal Computation and the Emergence of Mind," *Evolution and Cognition* 2, no. 1: 2–15.

37. Philip Kitcher, *Living with Darwin: Evolution, Design and the Future of Faith* (Oxford: Oxford University Press, 2006).

38. H. Allen Orr, "A Religion for Darwinians?" *New York Review of Books* 54, no. 13 (2007): 33–35.

6

Rapid Advances in Human Brain–Machine Interfacing: Ethical and Social Implications

Michael Bess

> Four legs and two voices—a most delicate monster!
>
> —Stephano to Caliban, *The Tempest*

Functional Convergence in Neuroscience, Informatics, and the Humanities

"REDUCTIONISM IS LIKE CHOLESTEROL," writes the psychologist Steven Pinker in *The Blank Slate*: "It comes in good and bad forms."[1] Bad reductionism, for Pinker, aims at collapsing the differences among distinct levels of reality: it consigns most features of the world to the status of mere epiphenomena and leaves a single aspect, such as physics or economics, dominant in the causal determination of events. An example would be the attempt to explain human love affairs by reducing them to expressions of *mere* economic self-interest or the *mere* transmission of genetic information through time.

Pinker has no truck with this kind of simplistic reductionism. He acknowledges that the reductionist impulse is apparently still tempting to some of his colleagues in the natural sciences (and also to some of his social-scientific colleagues who have chips on their shoulders about being second-class thinkers in comparison to the "hard" scientists). But overall, Pinker dismisses this kind of reductionism as an intellectually outmoded way of thinking. He cites the philosopher Hilary Putnam's deceptively simple example of an irreducible ordering into discrete levels in the natural world: the fact that a square peg cannot fit into a round hole.[2] This fact cannot be understood at the atomic level alone: even a complete account of the square and round objects, based on their atomic compositions, cannot predict the incommensurability of their two shapes. For that we need properties of molecular rigidity and geometric compatibility that only manifest at a qualitatively higher level. Thus, even the properties of inanimate objects require explanatory models more sophisticated than mere reductionism allows. A fortiori, of course, we won't get far in explaining the causes of World War I if we try to deduce them from the interactions of quarks and strings—despite the fact that quarks and strings indubitably constituted the raw material out of which the history of the year 1914 was composed.

Good reductionism, by contrast, is the wave of the future for Pinker: he notes that it is sometimes called *hierarchical reductionism* and "consists not of *replacing* one field of knowledge with another but of *connecting* or *unifying* them."[3] Instead of narrowing our understanding, it broadens it by showing the rich linkages and resonances among multiple layers of reality that we usually tend to consider causally separate. As an example of good reductionism at work, Pinker (an expert in cognitive linguistics) offers the following multilayered description of the language phenomenon:

> Language is based on a combinatorial grammar designed to communicate an unlimited number of thoughts. It is utilized by people in real time via an interplay of memory lookup and rule application. It is implemented in a network of regions in the center of the left cerebral hemisphere that must coordinate memory, planning, word meaning, and grammar. It develops in the first three years of life in a sequence from babbling to words to word combinations, including errors in which rules may be overapplied. It evolved through modifications of a vocal tract and brain circuitry that had other uses in earlier primates, because modifications allowed our ancestors to prosper in a socially interconnected, knowledge-rich lifestyle. None of these levels can be replaced by any of the others, but none can be fully understood in isolation from the others.[4]

One might quibble with Pinker's terminology here: to speak of *good reductionism* in this case is misleading, because no reduction is taking place at all in the description of language he offers. No level is being relegated to the status of a mere epiphenomenon. What we have instead is a webbing of interconnected but qualitatively distinct elements, each vitally important to the whole, coming together simultaneously in an explanation of how linguistic behavior occurs: overall purpose and functionality, real-time operation, neural basis, developmental stages in the organism, evolutionary history, and social context. A better term for this kind of multilayered etiology would be *functional convergence*, the irreducible complexity of all the ongoing causal interactions (some unidirectional, some bi- or multidirectional) among analytically distinct levels of order in a physically based phenomenon.

The cognitive scientist Douglas Hofstadter, in his recent book *I Am a Strange Loop*, gives an eloquent description of this variety of nondualistic yet also nonreductionistic etiology as it applies to the phenomenon of the sense of selfhood arising in a human brain:

> Seen at its highest, most collective, level, a brain is quintessentially animate and conscious. But as one gradually descends, structure by structure, from cerebrum to cortex to column to cell to cytoplasm to protein to peptide to particle, one loses the sense of animacy more and more until, at the lowest levels, it has vanished entirely. . . . A nondualistic view of the world can thus include animate entities perfectly easily, as long as different levels of description are recognized as valid. Animate entities are those that, at some level of description, manifest a certain type of loopy pattern, which inevitably starts to take form if a system with the inherent capacity of perceptually filtering the world into discrete categories vigorously expands its repertoire of categories ever more toward the abstract. This pattern reaches full bloom when there comes to be a deeply entrenched self-representation—a story told by the entity to itself—in which the entity's *I* plays the starring role, as a unitary causal agent driven by a set of desires. More precisely, an entity is animate to the degree that such a loopy *I* pattern comes into existence, since the pattern's presence is by no means an all-or-nothing affair. Thus, to the extent that there is an *I* pattern in a given substrate, there is animacy, and where there is no such pattern, the entity is inanimate.[5]

The key phrase here, for our purposes, is *as long as different levels of description are recognized as valid*. Hofstadter takes care to acknowledge that each level has a distinctive and essential role to play; yet at the same time he is envisioning all the levels as nested and interacting features of a single homogeneous phenomenon. This delicate equipoise between a hierarchical ordering of qualitatively discrete functions, operating within the broader embrace of an ontologically unitary realm, is a perfect exemplar of functional convergence at work.

In this chapter I focus on another particularly rich example of functional convergence: the technologies of human brain–machine interfacing. These technologies exemplify functional convergence because they reflect a powerful nexus of advances in psychology, neuroscience, prosthetics, and informatics, as well as the dramatic implications of that nexus for our social and political institutions, and for our sense of what it means to be human. They take us all the way from potassium channels and long-term potentiation in a synapse, through the circuitry of brain tissue, to the mental properties illuminated by cognitive psychology, to the capabilities of implanted machines, to the informatic structures designed to interact functionally with the human sensorium, to the practical impact of that implanted machine in a single person's life, all the way up to the social and moral implications of a civilization in which such implants become commonplace. This is multilayered etiology par excellence.

Human beings, I will argue, are in the early stages of a process that involves incrementally merging with their machines. The process is accelerating, both as we learn more about human bodies and minds, and as we learn more about building machines that can interact functionally with human bodies and minds. Over the coming century, as the number of functional connections between humans and machines increases, as the potency of those connections deepens, as the number of individuals who incorporate such connections grows, our civilization will experience dramatic changes that will probably destabilize our social and political institutions in profound ways. It will constitute one of the greatest technological watersheds in human history.[6] We are dangerously unprepared, I will argue, for the epochal transformation that awaits us, down this road along which we are already advancing at an ever more rapid pace.

Toward a Civilization of Cyborg Technologies

From the Bonic Woman to the Six-Million-Dollar Man to the figures like Robocop and Darth Vader in comic books and movies, science fiction has familiarized contemporary culture with the imagery of human-machine convergence.[7] The term *cyborg*, short for *cybernetic organism*, was originally coined in the 1960s by NASA scientists envisioning ways to retrofit humans for space travel: it meant, in layman's terms, a sentient being that is a self-regulating integration of artificial and natural systems. The term has now come into common parlance, not through the achievements of the space program but through the far greater culture-shaping prowess of Hollywood.

And yet this link with science fiction is potentially misleading. The movies and novels of the sci-fi genre have taught us to place cyborgs within the (usually) outlandish context of the distant future. Because of this association, we are likely to overlook the fact that powerful technologies of human bioinformatic enhancement are actually coming into being all around us, in the familiar world of the present. When we read about some remarkable breakthrough in robotics, prosthetics, neuroscience, psychopharmacology, brain-machine interfaces, or cognitive psychology, we still tend to think of each development as an isolated "futuristic" anomaly occurring in an otherwise unaltered technological landscape. But this, I argue, is an inaccurate understanding of today's reality. Once we connect the dots, and begin grasping the cumulative importance of developments in all these fields, *taken together*, we begin to see that many technological capabilities long associated with the classics of science fiction are in fact becoming—gradually, incrementally—very real features of our contemporary world.

I refer to these new technologies of bioinformatic enhancement with the imprecise but evocative term *cyborg technologies*. The term has two disadvantages: it lumps together many discrete and highly diverse machine-organism pairings under a single category, and it tends to convey a rather sensationalized connotation. But these disadvantages are outweighed, in my view, by the fact that this terminology keeps us focused on the big picture: it transcends all the individual applications of science and technology in this domain and emphasizes instead the overall social outcome toward which their conjoined trajectory seems to be pointing. It keeps us acutely aware of the kind of lifeworld we are poised to create for ourselves—the deeper ethical and human consequences of our technological innovations. If it is indeed true that our society is heading toward an incremental convergence of humans and machines, then we have every right to use sensational language. This is very exciting, and very scary, stuff.

It is important to have a concrete sense of the kinds of machine-organism pairings we are actually talking about. To this end, I will give three examples of these pairings that actually exist today and and more examples of pairings that are under development for possible deployment during the coming decades.

Today

- A man named Johnny Ray collapses during a severe stroke and wakes up with a permanent case of locked-in syndrome. He is fully conscious but cannot move any part of his body except his eyelids. After all other avenues of healing are exhausted without success, researchers implant a single electrode, wired to a nearby computer, into the motor cortex of Ray's brain. After weeks of training and hard work, Ray is able to will the cursor on the computer screen to move around, merely by thinking "up," "down," "left," "right." Over several months, his dexterity vastly improves, and the ability to manipulate the cursor becomes second nature to him. He uses the machine to type out his thoughts, wishes, and feelings. He has escaped from his nightmarish isolation and returned to the rich world of language. When asked what it feels like to move the cursor, Ray types out N-O-T-H-I-N-G.[8]
- An owl monkey at Duke University has been fitted with a motor-cortex implant similar to Johnny Ray's, but possessing seven hundred electrodes and a high-bandwidth interface. The animal learns quickly.

After several weeks of training, it sits in front of a video screen, its arms dangling motionless at its sides. Using thought alone, the monkey controls a robotic arm in an adjoining room: it plays video games by manipulating a joy stick. The monkey appears to have incorporated the third arm into its sense of its own body just as seamlessly as Ray had done with the computer cursor. In a subsequent experiment, the link between monkey and mechanical limb is extended over the Internet: in real time, from a lab at Duke, the monkey manipulates a robotic arm at MIT.[9]

• A blind man named Jens Naumann is outfitted with a special pair of glasses that have a tiny video camera mounted on them. The camera feeds its data to a computer on Naumann's belt; the computer translates the video input into neural signals modeled on those that a healthy eye sends down the optic nerve to the brain. Through an implant into Naumann's visual cortex, the computer sends real-time images from the camera directly into his brain. Result: grainy, blurred vision projected in a crude dot-matrix pattern, but vision nonetheless. Naumann can see well enough to drive a car around a parking lot without hitting streetlights or other cars. "It was a very nice feeling," he reports.[10]

Two to Five Decades in the Future

Between 2001 and 2004, the National Science Foundation convened a series of large interdisciplinary conferences to chart the landscape of new enhancement technologies.[11] At the heart of these meetings lay the acronym NBIC, the convergence of nanotechnology, biotechnology, informatics, and cognitive science in an accelerating series of synergistic innovations for improving human performance. The assembled scientists, business leaders, medical doctors, humanists, and technologists were asked to describe the state of the art in their respective fields when it came to human enhancement and to hazard their best guesses of where the technologies would be by the year 2050. It is worth emphasizing that these people had been invited by the NSF because they were top professionals in their areas—people primarily interested in establishing better cooperation among universities, businesses, and government in speeding along innovation. This was not, in other words, a fringe gathering of wild-eyed futurists indulging a penchant for flamboyant prophecy. Here (in their own words) are some of their predictions:

• Fast, broadband interfaces between the human brain and machines will transform work in factories, control automobiles, ensure military superiority, and enable new sports, art forms, and modes of interaction between people.

- Memory enhancement will improve human cognition by such means as external electronic storage and infusion of nerve-growth factors into the brain.
- Many people will carry with them a highly personalized computer database system that understands the user's emotions and functions as an artificial-intelligence advisor to help the person understand their own feelings and decision options.
- Computer-generated virtual environments will be so well tailored to the human senses that people will be as comfortable in virtual reality as in reality itself.
- Devices directly connected to the nervous system will significantly enhance human sensory, motor, and cognitive performance.
- Neuroceuticals—nonaddictive neurochemical brain modulators with high efficacy and negligible side effects—will cure mental illness and expand artistic expression.
- Nanoenabled sensors, implanted inside the human body, will monitor metabolism and health, diagnosing any health problem before the person even notices the first symptom.
- A combination of techniques will largely nullify the constraints associated with a human's inherent ability to assimilate information.
- Convergence of technologies will be so central to social change that it will transform the law and legal institutions, as various legal specialties merge, new ones emerge, and new issues challenge courts and legislatures.
- War-fighters will have the ability to control vehicles, weapons, and other combat systems instantly, merely by thinking the commands or even before fully forming the commands in their minds.[12]

It behooves us to adopt a resolutely skeptical and cautious approach to these kinds of predictions, for humans are notoriously bad at knowing what the technological and social future holds—and expertise in a given field seems to correlate fairly poorly with accuracy in making long-term predictions for that field. Sometimes the experts err wildly on the side of optimism, as in the case of the widespread 1950s assumptions that flying cars and unlimited free energy would be prominent features of the year 2000. Sometimes they err the other way, as in the case of the physicist Lord Kelvin's 1895 pronouncement that "heavier-than-air flying machines are impossible"—or (another physicist-Lord) Ernest Rutherford's dismissal of the idea of nuclear fission in 1934 as "the merest moonshine."[13] There is no doubt that many serious setbacks (most of them unforeseeable at present) will punctuate the advance toward the world depicted by the NSF's conclave of experts. The society of the year 2050 is likely to be quite different from anything we can imagine today: some aspects of our social world will prove to possess far greater staying power than we thought;

other aspects that we assumed were stable will no doubt change in completely unexpected ways.

Having said this, however, the future depicted by the NSF experts should give us pause. Most of their predictions are supported by a rapidly growing literature on human enhancement published in the past ten years—a wide-ranging literature that is both methodologically and ideologically diverse but that converges toward one unequivocal conclusion: partially redesigning human capabilities will increasingly become technologically feasible over the coming decades.[14] For the purposes of the present chapter, I am focusing only on cyborg technologies, but it is worth underscoring that the advent of these technologies is likely to be paired with two other, equally potent, categories of human enhancement: pharmaceutical boosting and genetic intervention.[15] We do not have space here to consider these two other domains of enhancement science, but throughout the present discussion we need to bear in mind that many cyborg technologies may well be operating *in synergy with* profound chemical and genetic modifications of the human constitution.[16] The upshot is straightforward: even if only a portion of what these experts envision becomes a reality, and even if the development of such capabilities takes twice as long as they think, we are still staring in the face of one of the great disjunctions in human history.

There are two main levels (analytically distinct, though richly interconnected in practice) in which the advent of these cyborg technologies will pose profound ethical challenges: the level of the individual and that of the broader society as a whole.

Cyborg Individuals: Three Questions

Is the Distinctiveness of Humans Doomed?

The scientific fields of robotics and artificial intelligence got off to a highly promising and ebullient start in the 1950s, then went through a somewhat mortified phase of unexpected difficulties and setbacks in the 1970s and 1980s.[17] Today these fields are once again expanding rapidly, both in the number of labs (academic and corporate) devoted to them and in the sheer variety of their basic design approaches. The explanation for this six-decade trajectory—marked by an initial surge, followed by a middle period of doldrums, and finally culminating in renewed vigor and creativity since the early 1990s—is not difficult to discern. It lies in the assumption made by artificial-intelligence and robotics researchers in the 1950s that their machine creations would face their easiest challenges in behaviors that humans find easy and their hardest challenges in behaviors that humans find difficult. In actuality, however, they gradually came to a collective realization that something like the opposite was the case: walking across the house to fetch a beer from the fridge, something that humans find easy (perhaps excessively so), turned out to be devilishly hard to program in a machine. But advanced computation, mathematical modeling, and even sophisticated games of strategy like chess, which humans prided themselves in considering distinctively tough challenges, turned out to be not exactly simple but also not nearly as difficult for machines to accomplish as their designers had assumed.

At the risk of oversimplifying a rich and highly diversified field of inquiry, we can say that two basic schools of thought emerged over the years following the 1960s. One school, whose thinking might be called the top-down approach, clung tenaciously to the idea that a successful robot or artificial intelligence must be able to model the world around it, in a form of representation patterned after human conscious awareness. The key goal for these researchers, such as Marvin Minsky or Douglas Lenat, therefore lay in the arduous task of constructing such a digital model of the world inside their machines. They have enjoyed some notable successes, but so far their massive efforts have failed to come close to achieving a level of functionality that would satisfy the purchaser of even the most rudimentary household robot helper. Thus far, the top-downers have had a frustrating journey.

Perhaps inspired by this intractable record of setbacks and slow progress, another school (their most famous member being MIT's Rodney Brooks) began in the early 1980s to develop what might be called the bottom-up approach. The short-term goals of these researchers were, on the surface, deceptively modest: they sought to build small embodied and situated machines that could move about, navigate the obstacles in a room, sense their environments, and perform elementary tasks, all without using any higher form of representation but rather relying on very simple and iterative behaviors more akin to reflexes and conditioned responses. Nonetheless, this apparent simplicity of design belied the long-term aims of the bottom-up school: they believed that they were replicating, within the machine world, the steady upward march of behaviors achieved in the animal world by eons of evolution. Amoeba, insect, dog, monkey, human: following a similar path of ascending sophistication, each level building on its predecessor, the machines would slowly ascend their own Great Chain of Being. The machines would start simple and humble, getting very good at doing the basics; then they would move up a level and master more complex tasks; and this process, repeated through many generations of machines, would ultimately begin to produce truly human-like behaviors.[18]

Today, both these schools of thought are still feverishly at work, and it is perhaps even misleading to think of them as rivals any more, since they often borrow from each other's technical achievements and use amalgamated or overlapping concepts and devices. One result of their recent success has been an increasingly widespread questioning of the distinctiveness of humans themselves. This is nothing new in human history, of course: Descartes serenely used mechanistic metaphors in conceptualizing the mind-body relationship, and many nineteenth-century philosophers drew heavily from concepts of machinery, hydraulics, and electric circuitry to describe the phenomena of biology.[19] Nevertheless, this kind of metaphorical thinking has arguably risen to a qualitatively new level in the decades since 1945, to the point where, in some instances, the metaphors cease to be construed as metaphors anymore and take on an unprecedented literalness. In many accounts (both in academic discussions and in the popular media), humans emerge as organic machines, possessing hardware and software attributes just like any computer (only far more sophisticated, to be sure). Over the years, as the machines become increasingly ubiquitous, and some of them gradually acquire capabilities to emulate more and more of our behaviors, we humans increasingly seem to be framing our own self-understanding—our very identity—as if we ourselves were, in the final analysis, nothing more than highly complex machines.[20]

Here, therefore, we return to Steven Pinker's distinction between good and bad reductionism. If the *nothing more* in the preceding sentence means that our psychological, emotional, and moral experiences are completely explicable at the level of atoms and molecules, then it becomes hard to frame any moral position at all regarding the implications of cyborg technologies. After all, if we are ourselves mere machines made of biological matter, then what fundamental difference is there between enhancing our own capabilities through prostheses and implants and upgrading a computer or modifying a household appliance? The disciplines of physics, chemistry, and biology are not equipped to answer this question.

If, however, the *nothing more* refers to the kind of functional convergence described earlier in this chapter, then we are really saying that humans operate on many qualitatively distinct levels simultaneously: atoms, cells, organ systems, organisms, social groups, and networks of social groups. If this is the case, then we can call on the appropriate level of etiology to address the moral questions: is it right to enhance a human? Are some modifications morally preferable to others? Can a cyborg accumulate so many profound modifications that he or she ceases to be classifiable as a human person at all? These are examples of questions that require the etiological levels of human social existence—ethics, history, anthropology, psychology, and theology, for example—to address coherently.

Is the distinctiveness of humans doomed, then? The concept of functional convergence can help us to address this question quite concretely. The answer is probably yes, if human distinctiveness requires us to posit some kind of soul or other mysterious quality that transcends empirical observation. This kind of dualistic understanding of our status in the world seems to be completely unsupported by the impressive accumulation of scientific evidence thus far.[21] Only religious faith can support a continued adherence to this kind of human "specialness."

The answer is no, however, if we are content to interpret distinctiveness more modestly. A nondualistic perspective like Pinker's or Hofstadter's still allows for an important measure of human "specialness." It is a quality derived from the unique set of physical and mental capabilities that characterizes our species: the ability to use reason, language, tools, and social networks in a way that gives rise to cultural meaning and a scientific understanding of the world around us. No other species can do this: it is a capability that radically separates humans from all other creatures. At the same time, nothing about this capability requires metaphysical additions in order to be understood—and increasingly well understood, we might note—by the convergent labors of scientists, humanists, and social scientists following purely empirical methods.[22]

Are Human Individuals Undergoing a Gradual Process of Commodification?

Up until this point in history, people could undergo change in two basic ways: they could alter their minds and character over time, through learning, maturation, and the acquisition of experience, and they could alter their physical bodies, through embellishments, surgery, or medication. In most cases (with a few notable exceptions, such as a lobotomy or a sex change), such modifications were generally considered nonthreatening to the basic identity of the person being altered. The underlying constitution of the individual in question remained stable, while undergoing incremental changes that left that individual's overall sense of personal continuity and selfhood intact.

With the powerful biological and electromechanical interventions now opening up before us, this situation is set to change dramatically. For the first time, it is the very constitution itself of a person that will be subject to radical shaping or transformation. At the end of the process, one potential outcome is that a qualitatively different person will emerge, someone who looks, behaves, and feels completely differently from the person who would have existed in the absence of the intervention. Such a person would unavoidably face a host of problems that currently do not characterize the human world but solely the realm of machines, devices, artifacts. People, in other words, would suddenly find themselves encountering the problems of products.

To cite only two examples of such problems, among the many possible ones:

The Problem of Ownership

Companies or agencies that go to the trouble of developing the enormously complex and expensive processes of human redesign would undoubtedly want to protect their property rights through patents. Already today we see considerable debate and litigation over the early stages of the patenting of human genetic material: it is not hard to imagine what a complex tangle of legal issues would surround such matters in fifty years. If a cyborg person decides to tinker with the package of installed features after they have been incorporated into his or her body, can a company sue to block this from happening? Who is liable if the design fails? And at a more ethereal but deeply important level: what effects would it have on human identity for certain core aspects of an individuals being to be patented?[23]

The Problem of Obsolescence

Presumably, the science and technology of human redesign would get better and better as time goes by, and each new generation of enhancement packages would offer safer, more reliable, and markedly improved features. Would the resulting generations of redesigned humans look back on older generations as outdated models? Would people be forced, whether they liked it or not, into an endless spiral of enhancement updates and prosthetic upgrades—a relentless sort of "arms race" (pardon the pun)? Would humans become, in effect, akin to the Microsoft Windows software: Humans '95, Humans '98, Humans XP, Humans Vista? Each generation noticeably superior (at least by one set of measures) to those that came before, each generation surpassed in turn by the following one?[24]

Clearly, any technology that tempts us to think of people increasingly in the same terms as artifacts poses a serious risk to human dignity. By this usage of the term *dignity* we usually mean a quality of intrinsic and absolute value that all humans share equally, whether a tiny baby, a great genius, or a mentally handicapped person.[25] It is this quality that makes us say, for example, that the crime of murder is always equally wrong, regardless of the victim's personal characteristics.

We are not speaking here merely of what might be called the Snob Problem, the tendency one individual might have to be snooty toward another whose enhancement package was less potent or came from a less-prestigious manufacturer. The deeper challenge will lie in how the citizens of such a society might come, over time, to devalue human personhood in general. To view an individual as a product (or even as a partial product) would be to dehumanize that individual—and probably by extension all individuals—in a fundamental way.

This is best illustrated by the Humans '98, Humans XP example given above. Suppose Mary Smith has just undergone an upgrade from Human Boosting Package '98 to Package XP, whereas her neighbor, John Jones, is still operating with Package '98. They get together after her return from the installation clinic and compare levels of functionality: her manual dexterity, speed in reading, comprehension of abstract concepts, memory retention, and mathematical skills are all significantly better than his. Along comes another neighbor, Steve Brown, who is still using Human Boosting Package '95. The three of them chat for a while. Steve marvels at Mary's new capabilities and then walks on. It is clear to both Mary and John, as they watch Steve walking away, that the fellow was simply unable to grasp the full extent of Mary's new powers: the subtlety and richness of her enhancement package lay too far beyond his reach.

It is hard to imagine that, at some level, these three individuals would not be tacitly classifying each other along a gradient corresponding to their different capabilities. They might all still insist that they viewed each other as possessing equal human dignity and as embodying the full measure of worth accorded to any human. But the technology would pull strongly in the opposite direction: over time many people in such a society would be likely (whether consciously or not) to blur the distinction between measurable attributes of functionality and fundamental human worth. To the extent that this happened, human persons (at all levels of functionality, high and low) would be reduced to the level of mere things. This, indeed, is one of the more striking features captured by Aldous Huxley in *Brave New World*: it is not just the lower-caste Epsilons and Gammas who are degraded by the hierarchical division of labor. All persons, including the upper-crust Alphas and Betas, are fundamentally dehumanized by this system, because it reduces human worth to social utility rather than relying on the intrinsic value of individual human dignity.[26]

Apart from the inherent degradation that such a functionally based system of worth entails, it is also repugnant for more practical reasons: it opens up humans for treatment as mere instruments or tools.[27] Two twentieth-century ideologies that explicitly subscribed to this kind of instrumentalization of humans were Stalinism and Nazism; in both cases, human individuals were understood as subordinate in value to the higher moral worth of the collective entity, the proletarian class or the Aryan race. Though individuals in these systems did possess some intrinsic worth, they could still be legitimately sacrificed in the name of the higher-level identity of the collective whole. Insofar as cyborg technologies promote this kind of instrumental understanding of human worth, they pose a grievous threat to the most fundamental values on which our civilization is founded. A central moral challenge of the coming decades will be to prevent the technologies of enhancement from eroding the foundations of equality and human dignity on which our democratic political and social systems rest.

Where Do the Boundaries of Personal Identity Lie for a Human Cyborg?

Let us return for a moment to one of the predictions offered by the NSF experts: "Fast, broadband interfaces between the human brain and machines will transform work in factories, control automobiles, ensure military superiority, and enable new sports, art forms, and modes of interaction between people."

If a person can connect directly to an informatic device over a broadband interface, then that same person can presumably also connect *through* the informatic device to another person on the other end. The implications are extensive. People could communicate with each other directly, brain to brain, in the same way as computers today connect directly with each other over the Web. Individuals could share thoughts and memories, not by using words to describe events that had happened to them, but by transmitting the neural encodings of those mental representations straight from brain to brain.[28] I would be able to experience your taste of the raspberry croissant exactly as you have stored it in your own mind.

Well, not *exactly*. It is likely that my second-hand experience of your memory of eating the croissant will still differ significantly from your own first-hand experience, because of the intrinsically interconnected nature of all memories. A person's mental representations of past experiences probably do not exist as discrete packages of information, akin to film clips stored in a computer or on a DVD.[29] Rather, the evidence available today from psychology and neurophysiology suggests that particular memories exist as embedded elements in a vast webbing of associative connections, with distinct aspects of each experience linked to similar or associatively conjoined aspects of other experiences. This associative mechanism appears to underlie the basic process through which abstract categories are formed in the mind.[30] Each supposedly discrete memory is actually surrounded by an aura of concentric circles of associations—primary associations, secondary associations linked to the primary ones, tertiary associations linked to the secondary ones, and so on—extending outward like ripples in a pond of meaning.[31]

Therefore, when you upload to me your particular set of neural encodings representing the croissant experience, the electronic device carrying out the transmission of information is presumably having to select only the most immediate collection of neural patterns and associations directly linked to the experience itself. The alternative would be to transmit an infinitely recursive set of associations surrounding the event in your mind: an unlimited regression until every association in your mind would be implicated. This would of course constitute an extremely large "file," and sending such a file would be either impossible or impractical.

Consequently, the transmission of memories directly from brain to brain is likely to involve an initial act of selection, before transmission, followed by an unavoidable alteration of content, as the neural encodings are recontextualized within my own brain upon arrival. I would link the incoming croissant flavors to my own associations with raspberries—with France, with breakfasts enjoyed with my mom back in Marin County, California, in 1968—and all this would give the experience a decidedly different "flavor" from the one experienced by you.

Nevertheless, even with this important caveat, the fact remains that a direct transmission of mental representations from one brain to another would constitute a real qualitative leap. Nothing remotely like it has ever occurred in human history—or in the entire history of the natural world on earth. A recontextualized memory of your croissant experience, presenting itself in my mind as a transmission from you, would bring me closer to experiencing the world from your perspective than any other kind of sharing in the history of human culture. I would be able to know, in an unprecedentedly intimate way, what you tasted, what you perceived, and perhaps (depending on the amount of associative aura included in the transmission) a great deal of what you thought and felt while you chewed. (One can already imagine TV ads of the 2050s eagerly touting the advantages enjoyed by customers who have upgraded their brain-to-brain transmission devices from Version 7.3, with "normal resolution," to the new and improved Version 8.0, with "high definition" capability. "Up to four times more aura than Version 7.3. Now you can *really* know what she was feeling!")

The implications of such a technological capability for human personal identity are profound. Memories, and the feeling of those memories being "mine," are central to our sense of who we are as individuals. It is hard to see, therefore, how such an unmediated sharing of experiences would not profoundly destabilize the boundaries of selfhood. Which memories are truly mine, and which are yours? Even if I have taken care to tag the memories, upon receipt, as "not mine," it is likely that a prolonged and extensive sharing of intimate, rich, subjective experiences would eventually cause a blurring of the boundaries between me and you.[32]

This would be a wonderful thing for some purposes. I would be able to empathize with others in an extremely vivid way. Two lovers or friends could share their experiences and feelings with unprecedented potency. As a teacher, I could communicate with my students far more effectively, directly transmitting my own images, understandings, ways of seeing and thinking. And my students could transmit right back to me: their own perplexities, reactions, misgivings, and distinctive ideas would appear in full living color (and taste and touch) within my own mind.[33]

But the downside is also equally potent. It would be hard, in such an unmediated world of experiential overlapping, to know where I end and others begin. If I can experience the feeling of skydiving, in all its glorious richness—physical, perceptual, emotional—simply by downloading your memories of skydiving yesterday morning, then does this not radically subvert the distinction between real experience and virtual experience? Why should I undergo the risks and costs of experiencing something like skydiving when I can simply link up with someone who has done so and download the experience directly? And, of course, the same technologies that enabled this kind of person-to-person sharing would presumably also enable me to send and receive an infinity of such experiences over the Web. I could purchase, trade, simply give away, or freely take in just about any experience recorded by any connected human brain anywhere in the world. Wow.

The implications for personal privacy are equally dramatic. If the police or the government (or my insurance company) needs to know what I'm thinking, can I prevent them from linking into my brain and simply uploading the contents of my conscious (and perhaps unconscious) awareness? Can my mind be scanned against my will?[34]

Admittedly, this scenario certainly does sound like far-fetched science fiction. But it is a scenario that follows as a direct implication of precisely those technologies that the NSF's assembled scientists and technologists are envisioning as plausible a mere half-century from now. This much is undeniable: our society today is devoting considerable resources to the development of broadband brain-machine interfaces. It is pouring even greater resources into the effort to record, interpret, and manipulate the neural encoding of memories and states of mind. Over the coming decades, to the extent that these capabilities become a reality, each of us will be forced to confront the radical moral issues I have been describing. The commodification of humans, the blurring of individual identities, the erosion of privacy—debates over these kinds of issues will almost certainly pervade our children's world.

A Cyborg Society: Three Questions

Will Cyborg Technologies Tend toward Conformist Homogenization?

Let us suppose that cyborg technologies, as they incrementally develop over coming decades, will become available for purchase on the open market (with some restrictions and regulation imposed by government agencies, as applies today to many medical or consumer products). One possible result would be a gradual homogenization of the human species, as individuals responded to advertising and social pressures by purchasing and installing very similar cyborg technologies for themselves and their families. Over time, the natural differences among people might tend to be overshadowed by the potent capabilities that they shared in common through acquired technologies.

In a hunter-gatherer society, for example, it matters a great deal whether I can see well and run fast, but in today's world the technologies of optometry and mechanized transportation allow me (a myopic, rather plodding fellow) to lead as full and rich a life as someone who naturally sees much better and runs much faster than I do. The technology levels the playing field and overrides the functional disparities bestowed on us as individuals by Mother Nature.

In a similar way, cyborg technologies might plausibly lead toward a further narrowing of the differences among people. Those with weak capabilities of mental recall would rely on memory amplifiers to make up for their deficiencies; those who scored lower on IQ tests would turn to various deductive- and analytical-boosting devices to help them participate as equals with other citizens. The result, played out on a macroscopic social scale, would be a narrowed, more homogeneous range of behaviors. Just as consumer fads in today's world lead to widespread homogeneity in the kinds of appliances we buy, foods we eat, movies we watch, and houses we live in, so the cyborg technologies would lead to similar convergences in the way we enhance our physical and mental capabilities. To someone who places a high intrinsic value on human diversity, this would amount to a serious impoverishment of our species.

Will Cyborg Technologies Tend toward Species Fragmentation?

My own sense, however, is that the foregoing scenario of homogenization is actually much less likely to come about than the exact opposite one: a scenario in which cyborg technologies push human diversity to the breaking point.[35] All human societies exist in a state of tension, or dynamic equilibrium, between the centripetal tendencies of social conformity and the centrifugal tendencies caused by the innate diversity of individuals and by the cultural divergences among social groups. Cyborg technologies would be likely, in my view, to strengthen the centrifugal tendencies in ways that are historically unprecedented. They would do so for two main reasons.

First, they would enormously exacerbate the age-old economic division between haves and have-nots.[36] Wealthy persons would be most likely to gain preferential access to the most potent enhancements, both for themselves and for their children. This would result in a vicious circle, in which the most highly enhanced individuals would be able to compete more effectively for society's resources and opportunities, thereby securing for themselves the greatest rewards, which they would in turn invest in still more potent enhancements. The relative gap between rich and poor would grow ever wider—and this time around it would be a gap inscribed not just in social status, influence, and wealth, but increasingly in measurable differentials of physical and mental capability. The richer citizens, in other words, would incrementally become faster, smarter, stronger, healthier, and better looking than their economically challenged compatriots.

Some defenders of cyborg technologies have acknowledged this inherent tendency toward radical inequality but have argued that resolute government policies to ensure equal access to these technologies would suffice to avert such a nasty outcome. If enhancement technologies were federally subsidized, they argue, so that all citizens could afford them, then the biopolarization of society need not occur.[37] This argument fails to persuade, however, for one important reason: all we need do is consider the profoundly unequal access to basic education and health care that pervades our current social order. Generations of Americans have struggled, and failed, to solve this problem: we have experienced utter defeat as a society in guaranteeing the most elementary equality in education and health care to our citizenry. If we cannot provide even rudimentary equality in these fundamental aspects of existence today, why should we believe that our society will have any greater success in doing so with enhancement technologies half a century from now?

A second reason for expecting greater heterogeneity among a cyborg citizenry has to do with human culture. A key impulse in most human beings— an impulse observable in virtually every society throughout history—has been to divide into groups and subgroups demarcated by highly charged cultural markers of difference. Race, ethnicity, religion, identification with locality, sexual orientation, role in the division of labor, gender, physical appearance, nationality, degree of wealth, caste, tribal affiliation, ideological preference— these are only a sampling of the myriad categories people spontaneously use to divide themselves from each other.[38] If we factor the cyborg technologies into this powerful cultural force field of difference creation, it is not hard to imagine the result. How many different ways can we be (or at least think of ourselves as being) faster, smarter, stronger, healthier, and better looking? In a market economy—even a highly regulated one—the technologies of human enhancement are likely to result in a free-for-all of self-modification by people seeking to mark their individuality through a unique cyborg configuration.

While it is true that consumer fads, advertising, and other social pressures will play a homogenizing role, I think their conformist impact will be vastly outweighed by the sheer power that the cyborg technologies will have to engender real and profound differences in the people who adopt them. These enhancement technologies, in other words, are not mere superficial markers of social difference, like clothes, jewelry, or tattoos. They hold the potential to alter the fundamental constitution of individuals—their overall profile of physical and mental capabilities.

Cyborg technologies, in short, could plausibly bring about a gradual fragmentation of our species into discrete subcultures and lineages that have increasingly little in common with each other. One can even imagine a full-scale evolution of *Homo sapiens* into a series of parallel successor species. Some writers (many of whom refer to themselves as *transhumanists*) have argued that this would not necessarily be a bad outcome, but merely another stage in the long, winding road of evolution.[39] They ask, Who are we to assert that today's particular configuration of human physical traits and mental abilities is the be-all and end-all of our collective destiny?

Nevertheless, whatever one makes of species fragmentation from a moral point of view, it is clearly bad news from the practical perspective of intergroup rivalries, civil strife, and perhaps even war. Will a population of highly enhanced humans be able to coexist peacefully alongside a population of unenhanced humans? And within the increasingly heterogeneous population of the enhanced, might we not expect all manner of rivalries, jealousies, and other forms of cultural hostility to arise? Human society, throughout its long history, has not established a very promising track record when it comes to dealing with fundamental diversity among individuals and groups. Au contraire: we humans have exhibited an unfortunate tendency to be suspicious of, hate, discriminate against, or (more expansively) enslave or exterminate those groups who looked or acted profoundly differently from ourselves. Cyborg technologies, in this sense, will pose an unprecedented challenge to our sense of civil and social solidarity. We will need to find deeply resonant and persuasive ways to perceive our fellow citizens as members of a common family of beings—despite their dramatic differences from us in appearance, behaviors, and capabilities.

As a historian, I cannot help but regard such morally idealistic imperatives with a strong measure of skepticism. On the other hand, one can argue that the rapid development of military technologies over the past century has exerted a parallel, and equally important, pressure nudging our species in the same direction. Either we learn how to get along, despite all our differences—developing effective tools and institutions for nonviolent conflict resolution—or we wind up as a dead-end sign on that long, winding evolutionary road.

Can Society Control the Development of Cyborg Technologies?

Suppose, for the sake of argument, that a majority of Americans today were to vote to enact a comprehensive ban on all human-enhancement technologies, including brain-machine interfaces and other such cyborg devices.[40] Could the development of these technologies be stopped? I doubt it. The reason is simple: cyborg technologies are inextricably interwoven with too many areas of scientific research and technological innovation that have become far too important for us to give up. A ban on such devices, if it were to be truly effective, would require us to curtail drastically much of the ongoing work being done in informatics, robotics, prosthetics, and significant areas of medicine—not to mention a good deal of the more fundamental research being done in neuroscience, cognitive psychology, artificial intelligence, biomedical engineering, and nanoscience. It would require such a ban to apply not just at the national level but simultaneously and equally effectively in all the world's countries (or else the researchers and funding would simply migrate overseas to the areas of lowest regulation). Such a coordinated global relinquishment almost certainly lies beyond the reach of any government or coalition of governments to impose.

This does not mean, however, that we are powerless, as individual citizens and as a society, to exert some measure of partial control over *how* enhancement technologies develop. We may not be able to stop the river from flowing, but we can probably influence where the water channel passes and how rapidly it runs down its course. How should we go about this? The discussion thus far in this chapter suggests three distinct levels of possible intervention.

Restricting Specific Technologies

If we look at the sources of funding for research on cyborg technologies, we find four main categories: medicine, informatics, robotics, and the military. All four of these areas are carried forward by strong social and economic forces: the imperatives of health, communication and computing, manufacturing, and security. One of the most striking features of cyborg technologies is how easily they spill over from one of these areas to another. For example, if I can place a brain implant in a paralyzed patient, allowing him to manipulate a computer cursor by thought alone (as in the case of Johnny Ray), then I can also use a similar technology, down the road, to let healthy people manipulate the controls of a military aircraft, also by using thought alone. The transition from medical to military application is virtually seamless. This is precisely why the military research agency, DARPA, has invested heavily in putatively medical research into brain-machine interfaces: the "healing" technologies that pulled Ray out of his isolation lead straight into multiple other applications in advanced communications and weapons systems.[41] Almost all cyborg technologies display this kind of fundamental flexibility: wherever we can restore a lost functionality through some kind of human-machine pairing, we can also, a little further down the technological road, boost that functionality to new levels beyond what was previously the norm. Enhancement, in other words, is written deeply into the very nature of most human-machine pairings: the more advanced the technologies effecting this linkage, the more advanced the possibilities for boosting human capabilities.

This suggests a very practical conclusion. If we want to control what kinds of cyborg devices get created, then the best way to do this is not to adopt a blanket restriction on broad categories of science and technology.[42] Any such sweeping restriction would also force the abandonment of research in areas that are highly valued by large numbers of citizens and would therefore become politically unsustainable. Rather, the only practical means of proceeding would be to monitor closely the entire field of emerging cyborg devices, while keeping a clear set of guidelines in mind (safety, human dignity, and so on). Then, one would need to identify a specific human-machine pairing that violated one of the guidelines or that demonstrably paved the way for such a violation. At this point, government regulation could come into play, in a highly focused way, restricting the development of that particular category of human-machine pairing. For example, we might decide that certain kinds of linkages between human bodies and powerful weaponry should be restricted or that pairings between humans and animals (through a machine) would violate our moral sentiments. In this way, on a case-by-case basis, we could require that any significant innovation in cyborg technologies be licensed and undergo prior screening by an agency that would assess the safety and moral issues presented by the device.

Legislation Aimed at Correcting Social Impacts

Regulating particular technologies, however, would in itself be insufficient. As we have seen, cyborg technologies would bring about—whether directly or indirectly—unprecedented social problems of a nature that would require systematic government intervention to correct. Among the key areas we have identified, the issues of equal access, privacy, and species fragmentation stand out as particularly salient. Citizens of the year 2050 will probably need to create a fundamentally new legal framework to deal with these problems.

The issue of equal access, for example, may well require some form of socialized distribution system for cyborg devices, perhaps modeled on the universal health-care system currently operating in Canada and many Western European democracies. As I noted above, this would require an expansion of governmental services and so-called entitlement programs that would be unprecedented in the history of American social policy. Nevertheless, it is hard to see how any other course of action would even begin to address the issue of equal access in an effective manner.[43]

The issue of privacy will force a continual trade-off between the inherent tendency of cyborg pairings to open up our minds to the world and the high value most people place on having a closed arena of intimacy that is theirs alone. Privacy legislation will have to reflect this trade-off, while also balancing the civil rights of citizens with the imperatives of national security. Similarly, the danger of species fragmentation may result in legislation that restricts the degree to which humans may modify their physical and mental constitutions: we can expect lively debates over the tension between the individual's "right to cyborg" and society's need to maintain a baseline level of species unity.

Social Programs Aimed at Building a New Civic Culture

Even this new legal framework, however, would still prove inadequate to meet the challenges posed by cyborg technologies. The grave problems of human commodification, and of extreme diversity among the citizenry, would not be satisfactorily addressed by legislation alone. Far-reaching new social programs would be needed to encourage the development of a new cultural ethos of unprecedented tolerance and solidarity among the citizenry of 2050. In order to counter the inherently dehumanizing aspects of human-machine pairings, this ethos would need to educate citizens to recognize absolute moral worth in all sentient beings, whatever their appearance or capabilities. It would need to instill a deep abhorrence for any value system that instrumentalized humans or treated them like mere objects. And through a value system that emphasized an equally strong sense of belonging to a more expansive human family, it would need to neutralize the strong divisive forces pulling apart a cyborg population. Educational systems and public media would need to be oriented toward the nurturing of this ethos, and the moral education given by parents to their children would need to prepare those young cyborgs for a more engaged and morally attuned citizenship.[44] A sense of deep respect for all manner of other beings, an ability to appreciate and enjoy extreme levels of diversity—these would need to become second nature to a new generation.

If this sounds utopian, that's because it *is*, in some measure, utopian. The irony of this situation is not unlike the one observed in the aftermath of Hiroshima by Albert Einstein, Leo Szilard, and other atomic scientists in their 1946 pamphlet *One World or None*.[45] Precisely because humans had attained an unprecedentedly high technological level, they now possessed the means to destroy their own civilization if they continued in their age-old traditions of warfare; therefore human beings urgently needed to make a commensurate leap into a new moral order, capable of resolving conflicts in more peaceful ways. Moral sophistication could not be allowed to lag too far behind technological sophistication. The value system espoused by moral leaders like Albert Schweitzer and Gandhi (and later by Mother Teresa and Martin Luther King Jr.) would need to become the beacon for coming generations: over the long haul, our very survival as a species depended on it.

The advent of cyborg technologies will present humans with a very similar challenge. Like the nuclear reactors of the mid-twentieth century, the cyborg devices of the mid-twenty-first century will probably come into our lives whether we as individuals approve of them or not. And like those potent machines of the last century, these new machines will simultaneously hold out great peril and hope. They may liberate people from many of their ancient afflictions and broaden the horizons of human creativity in extraordinary ways. Or they may result in catastrophe, through our inability to adapt to the powers they give us, and through our mismanagement of the radical transformation they will cause in our society.

Notes

I thank the National Human Genome Research Institute, Program on Ethical, Legal, and Social Implications, for its financial support of this project (Grant #RO3HG003298-01A1). For the research fund that has helped to support this work, my thanks also to the College of Arts and Science, Vanderbilt University, and the endowment for the Chancellor's Chair in History. I owe a special debt to Volney Gay and John McCarthy, of Vanderbilt's research seminar on "Scales and Hierarchies: Implications for Science and Religion," based in the Center for the Study of Religion and Culture, for their counsel and support, and to Vivian Ota-Wang, whose encouragement and intellectual guidance when she was at Vanderbilt's Center for Genetics and Health Policy proved invaluable in getting this project launched. I also thank the following persons whose willingness to offer practical counsel and scholarly expertise have greatly assisted my research: Andrea Baruchin, Ellen Clayton, Jay Clayton, Gerald Figal, Carol Freund, Michael Goldfarb, Kazuhiko Kawamura, Richard Alan

Peters, Matthew Ramsey, Susan Schoenbohm, Charles Scott, Arleen Tuchman, and David Wood. Finally, I thank my students in three honors seminars at Vanderbilt (in 2002, 2004, and 2007) for the wonderful intellectual exchange that has enriched my thinking on these matters.

1. Steven Pinker, *The Blank Slate: The Modern Denial of Human Nature* (New York: Penguin, 2002), 69. I have slightly altered the wording. The original reads, "Reductionism, like cholesterol, comes in good and bad forms."

2. Hilary Putnam, "Reductionism and the Nature of Psychology," *Cognition* 2 (1973): 131–46.

3. Pinker, *Blank Slate*, 70.

4. Pinker, *Blank Slate*, 70–71.

5. Douglas Hofstadter, *I Am a Strange Loop* (New York: Perseus, 2007), 359–60.

6. See, for example, Max Oelschlaeger, *The Idea of Wilderness: From Prehistory to the Age of Ecology* (New Haven, Conn.: Yale University Press, 1991); David Landes, *The Unbound Prometheus: Technological Change and Industrial Development in Western Europe from 1750 to the Present*, 2nd ed. (Cambridge: Cambridge University Press, 2003); Thomas Hughes, *Human-Built World: How to Think about Technology and Culture* (Chicago: University of Chicago Press, 2004).

7. Fredric Jameson, *Archaeologies of the Future: The Desire Called Utopia and Other Science Fictions* (New York: Verso, 2005); William Haney, *Cyberculture, Cyborgs, and Science Fiction: Consciousness and the Posthuman* (Amsterdam: Rodopi, 2006); Edward James and Farah Mendelsohn, eds., *The Cambridge Companion to Science Fiction* (Cambridge and New York: Cambridge University Press, 2003); Jutta Weldes, ed., *To Seek Out New Worlds: Exploring Links between Science Fiction and World Politics* (New York: Palgrave, 2003).

8. Ramez Naam, *More Than Human: Embracing the Promise of Biological Enhancement* (Louisville, Ky.: Broadway, 2005), 172–75, 218; John Hockenberry, "The Next Brainiacs," *Wired* 9, no. 8 (August 2001).

9. Naam, *More Than Human*, 179–81; Dennis Meredith, "Monkeys Consciously Control a Robot Arm Using Only Brain Signals; Appear to 'Assimilate' Arm as If It Were Their Own," in *DukeMedNews*, October 13, 2003, available online at: http://www.dukemednews.duke.edu/news/article.php?id=7100 (accessed November 14, 2008).

10. Lisa Cartwright and Brian Goldfarb, "On the Subject of Neural and Sensory Prostheses," in *The Prosthetic Impulse: From a Posthuman Present to a Biocultural Future*, Marquard Smith and Joanne Morra, eds. (Cambridge, Mass.: MIT Press, 2006), 142–46. Sanjay Gupta and Kristi Petersen, "Could Bionic Eye End Blindness?" CNN, June 13, 2002. Available online at http://archives.cnn.com/2002/HEALTH/06/13/bionic.eye/index.html (accessed November 14, 2008).

11. Mihail C. Roco and William Sims Bainbridge, eds., *Converging Technologies for Improving Human Performance: Nanotechnology, Biotechnology, Information Technology, and Cognitive Science* (New York: Kluwer, 2003); and *Managing Nano-bio-info-cogno Innovations: Converging Technologies in Society* (New York: Springer, 2006).

12. Bainbridge and Roco, eds., *Managing Nano-bio-info-cogno Innovations*, 339–44.

13. For the Kelvin reference, see Zack Lynch, "Neuropolicy (2005–2035)," in Bainbridge and Roco, *Managing Nano-bio-info-cogno Innovations*, 173; for the Rutherford reference, see Michael Bess, *Realism, Utopia, and the Mushroom Cloud* (Chicago: University of Chicago Press, 1993), 46.

14. Apart from the Roco and Bainbridge books, see, for example: Michael Chorost, *Rebuilt: How Becoming Part Computer Made Me More Human* (New York: Houghton Mifflin, 2005); Naam, *More Than Human*; Andy Clark, *Natural-Born Cyborg: Minds, Technologies, and the Future of Human Intelligence* (Oxford: Oxford University Press, 2003); Joel Garreau, *Radical Evolution: The Promise and Peril of Enhancing Our Minds, Our Bodies—and What It Means to Be Human* (New York: Doubleday, 2004); James Hughes, *Citizen Cyborg: Why Democratic Societies Must Respond to the Redesigned Human of the Future* (Cambridge, Mass.: Westview, 2004); Chris Gray, *Cyborg Citizen: Politics in the Posthuman Age* (New York: Routledge, 2002); Ray Kurzweil, *The Singularity Is Near: When Humans Transcend Biology* (New York: Viking, 2005); Steven Marcus, *Neuroethics: Mapping the Field* (New York: Dana Press, 2002); Douglas Mulhall, *Our Molecular Future: How Nanotechnology, Robotics, Genetics, and Artificial Intelligence Will Transform Our World* (Amherst, N.Y.: Prometheus, 2002); Steven Rose, *The Future of the Brain: The Promise and Perils of Tomorrow's Neuroscience* (Oxford: Oxford University Press, 2005); Marquard Smith and Joanne Morra, eds., *The Prosthetic Impulse: From a Posthuman Present to a Biocultural Future* (Cambridge, Mass.: MIT Press, 2006); Edward Tenner, *Our Own Devices: How Technology Remakes Humanity* (New York: Vintage, 2003); Bruce Mazlish, *The Fourth Discontinuity: The Coevolution of Humans and Machines* (New Haven, Conn.: Yale University Press, 1993); Eric Parens, ed., *Enhancing Human Traits: Ethical and Social Implications* (Washington, D.C.: Georgetown University Press, 1998); Gregory Stock, *Redesigning Humans: Choosing Our Genes, Changing Our Future* (New York: Mariner, 2003); Nicholas Agar, *Liberal Eugenics: In Defence of Human Enhancement* (Oxford: Blackwell, 2004).

15. On chemical enhancement, see Carl Elliott, *Better Than Well: American Medicine Meets the American Dream* (New York: Norton, 2003); Nathan Jendrick, *Dunks, Doubles, Doping: How Steroids Are Killing American Athletics* (New York: Lyons, 2006); Angela Schneider, ed., *Doping in Sport: Global Ethical Issues* (New York: Routledge, 2007); Claudio Tamburrini, *Genetic Technology and Sport: Ethical Questions* (New York: Routledge, 2005); Andy Miah, *Genetically Modified Athletes: Biomedical Ethics, Gene Doping and Sport* (New York: Routledge, 2004). The topic of genetic enhancement of human capabilities has generated equally as much literature as the cyborg technologies. A sampling: David DeGrazia, *Human Identity and Bioethics* (Cambridge: Cambridge University Press, 2005); Glenn McGee, *The Perfect Baby: Parenthood in the New World of Cloning and Genetics* (Lanham, Md.: Rowman & Littlefield, 2000); Lori Andrews, *Future Perfect: Confronting Decisions about Genetics* (New York: Columbia, 2001); Justine Burley and John Harris, eds., *A Companion to Genethics* (New York: Blackwell, 2002); Troy Duster, *Backdoor to Eugenics* (New York: Routledge, 2003); Verna Gehring, *Genetic Prospects: Essays on Biotechnology, Ethics, and Public Policy* (Lanham, Md.: Rowman & Littlefield, 2003); Roger Gosden, *Designing Babies: The Brave New World of Reproductive Technology* (East Lansing, Mich.: Freeman, 1999);

Herbert Gottweis, *Governing Molecules: The Discursive Politics of Genetic Engineering in Europe and the United States* (Cambridge, Mass.: MIT, 1998); John Harris, *Clones, Genes, and Immortality: Ethics and the Genetic Revolution* (Oxford: Oxford University Press, 1998); Leon Kass, *Human Cloning and Human Dignity: The Report of the President's Council on Bioethics* (New York: Public Affairs, 2002); Daniel Kevles, *In the Name of Eugenics: Genetics and the Uses of Human Heredity* (Cambridge, Mass.: Harvard, 2004); Richard Lewontin, *It Ain't Necessarily So: The Dream of the Human Genome and Other Illusions* (New York: New York Review of Books Press, 2000); James C. Peterson, *Genetic Turning Points: The Ethics of Human Genetic Intervention* (Winona Lake, Ind.: Eerdmans, 2001); Pete Shanks, *Human Genetic Engineering: A Guide for Activists, Skeptics, and the Very Perplexed* (New York: Nation Books, 2005); Lee M. Silver, *Remaking Eden: How Genetic Engineering Will Transform the American Family* (New York: Harper, 1998); Ian Wilmut, *After Dolly: The Uses and Misuses of Human Cloning* (New York: Norton, 2006). An older, but still valuable, study is Jonathan Glover's *What Sort of People Should There Be?* (New York: Penguin, 1984).

16. The synergistic relationship between these three domains forms the focus of a broader book project of mine from which this chapter is derived: "Icarus 2.0: Technology, Ethics, and the Quest to Build a Better Human." For a full description, and for a comprehensive bibliography on human enhancement, see http://www.vanderbilt .edu/historydept/michaelbess/Currentbookprojects (accessed November 14, 2008).

17. Rodney Brooks, *Flesh and Machines: How Robots Will Change Us* (New York: Pantheon, 2002); Rodney Brooks and Sandy Fritz, *Understanding Artificial Intelligence* (Warner Books, 2002); Maureen Caudill, *In Our Own Image: Building an Artificial Person* (Oxford: Oxford University Press, 1992); Alison Cawsey, *The Essence of Artificial Intelligence* (London: Prentice Hall Europe, 1998); Paul Ceruzzi, *A History of Modern Computing*, 2nd ed. (Cambridge, Mass.: MIT, 2003); Daniel Crevier, *AI: The Tumultuous History of the Search for Artificial Intelligence* (New York: BasicBooks, 1993); Hubert Dreyfus et al., *Mind over Machine* (New York: Simon & Schuster, 2000); Marvin Minsky, *The Emotion Machine: Commonsense Thinking, Artificial Intelligence, and the Future of the Human Mind* (New York: Simon & Schuster, 2006); Hans Moravec, *Mind Children: The Future of Robot and Human Intelligence* (Cambridge, Mass.: Harvard, 1988); Hans Moravec, *Robot: Mere Machine to Transcendent Mind* (Oxford: Oxford University Press, 1999); Geoff Simons, *Robots: The Quest for Living Machines* (New York: Cassell, 1992).

18. Brooks, *Flesh and Machines*.

19. Evelyn Fox Keller, *Making Sense of Life: Explaining Biological Development with Models, Metaphors, and Machines* (Cambridge, Mass.: Harvard University Press, 2002); Martin Kemp, *Seen/Unseen: Art, Science, and Intuition from Leonardo to the Hubble Telescope* (Oxford: Oxford University Press, 2006); Erica Fudge et al., eds., *At the Borders of the Human: Beasts, Bodies, and Natural Philosophy in the Early Modern Period* (New York: Palgrave, 1999).

20. Alan Wolfe, *The Human Difference: Animals, Computers, and the Necessity of Social Science* (Oakland: University of California Press, 1993).

21. Pinker does the most comprehensive job I have yet encountered of demolishing the dualistic position in the first five chapters of *The Blank Slate*.

22. Pinker (again!) presents a persuasive treatment of this subject in the second half of *The Blank Slate*.

23. For a sensitive fictional treatment of this kind of instrumentalization of human beings, see Kazuo Ishiguro, *Never Let Me Go* (New York: Knopf, 2005).

24. For a discussion of the ethical and legal aspects of the obsolescence of enhancement packages, seen from the perspective of a writer who enthusiastically supports human enhancement, see Stock, *Redesigning Humans*.

25. On the concept of human dignity, see, for example, Martha C. Nussbaum, *Frontiers of Justice: Disability, Nationality, Species Membership* (Cambridge, Mass.: Belknap, 2006); Glover, *What Sort of People*; Jurgen Habermas, *The Future of Human Nature* (Cambridge, Mass.: Polity, 2003); David Heyd, *Genethics: Moral Issues in the Creation of People* (Oakland: University of California Press, 1992); Amelie Rorty, *The Identities of Persons* (Oakland: University of California Press, 1976); Parens, *Enhancing Human Traits*; Walter Glannon, *Genes and Future People: Philosophical Issues in Human Genetics* (New York: Westview, 2002); Harold Baillie and Timothy Casey, eds., *Is Human Nature Obsolete? Genetics, Bioengineering, and the Future of the Human Condition* (Cambridge, Mass.: MIT, 2004); DeGrazia, *Human Identity*.

26. Aldous Huxley, *Brave New World* (New York: Harper, 1932).

27. Not surprisingly, a heated debate exists over whether the risk posed by enhancement to human dignity is insurmountable and unacceptably high or whether, with appropriate precautions, it is a risk that is manageable and acceptable, given the benefits that enhancement offers. For works that take the former position, see Bill McKibben, *Enough: Staying Human in an Engineered Age* (New York: Times Books, 2003); Francis Fukuyama, *Our Posthuman Future: Consequences of the Biotechnology Revolution* (New York: Farrar, Straus, 2002); and Leon Kass, *Life, Liberty and the Defense of Dignity: The Challenge for Bioethics* (New York: Encounter Books, 2002). For works that take the latter position, see Naam, *More Than Human*; Hughes, *Citizen Cyborg*; Stock, *Redesigning Humans*; and Garreau, *Radical Evolution*.

28. Clark, *Natural Born Cyborgs*, 127ff.; Naam, *More Than Human*, 176ff.

29. This is a common misconception promoted by Hollywood sci-fi films. The most recent example is *Paycheck*, a movie about a man who works on top-secret engineering projects and has his short-term memory erased each time a project is finished so that he cannot divulge the nature of the devices he has created. The memory-erasing process is depicted as a straightforward scrolling through a linear sequence of richly detailed scenes, each one presumably encapsulating one second of the man's conscious awareness. The erasure then consists of simply deleting a filmic time segment, just as if one were erasing a section of a video tape. For all our vast ignorance about how memory works, we can say one thing with near certainty: this depiction of memory erasure is, from a scientific perspective, utter balderdash.

30. Joe Z. Tsien, "The Memory Code: Learning to Read Minds by Understanding How Brains Store Experiences," *Scientific American* 297, no. 1 (July 2007): 52–59. See also Jeff Hawkins, *On Intelligence* (New York: Times Books, 2004), chap. 4; Antonio Damasio, *The Feeling of What Happens: Body and Emotion in the Making of Consciousness* (New York: Harcourt, 1999); and Christof Koch, *The Quest for Consciousness: A Neurobiological Approach* (Greenwood Village, Colo.: Roberts, 2004).

31. Douglas Hofstadter describes this phenomenon in an evocative chapter title: "The Blurry Glow of Human Identity" (*I Am a Strange Loop*, 259–74).

32. The robot designer Hans Moravec acknowledges, and celebrates, this likely outcome in his two visionary works, *Mind Children* and *Robot*.

33. Clark, *Natural-Born Cyborgs*, chap. 5.

34. See the discussion in Jonathan Moreno, *Mind Wars: Brain Research and National Defense* (New York: Dana, 2006), chap. 5. See also Wrye Sententia, "Cognitive Enhancement and the Neuroethics of Memory Drugs," in Bainbridge and Roco, eds., *Managing Nano-bio-info-cogno Innovations*, 165–66.

35. See the discussion in Hughes, *Citizen Cyborg*, chap. 11.

36. Fukuyama, *Our Posthuman Future*, chap. 1.

37. Hughes, *Citizen Cyborg*, chaps. 12–14; Naam, *More Than Human*, chap. 11

38. Clifford Geertz, *The Interpretation of Cultures* (New York: Basic, 2000); Pinker, *Blank Slate*, chaps. 16–20.

39. Naam, *More Than Human*; Clark, *Natural-Born Cyborgs*; Hughes, *Citizen Cyborg*; Kurzweil, *The Singularity Is Near*; Stock, *Redesigning Humans*; Agar, *Liberal Eugenics*.

40. Three thoughtful advocates of this position (albeit from sharply differing ideological backgrounds) are McKibben, *Enough*; Fukuyama, *Our Posthuman Future*; and Kass, *Life*.

41. Moreno, *Mind Wars*. DARPA stands for Defense Advanced Research Project Agency.

42. Fukuyama concurs on this point. See "The Political Control of Biotechnology," in *Our Posthuman Future*.

43. A recent comparative overview of the American and European contexts in this regard is Martin Gulliford, *Access to Health Care* (New York: Routledge, 2003).

44. See Michael Bess, *The Light-Green Society: Ecology and Technological Modernity in France, 1960–2000* (Chicago: University of Chicago Press, 2003). A suggestive example of such a deliberate shift in cultural ethos is provided by the history of the environmental movement in Western democracies. Over the course of five decades, from the movement's birth in the 1960s to today, green activists have succeeded in bringing about a partial reorientation of our society's value system regarding the environment, along with a partial restructuring of our economic, political, and legal institutions in this domain. The result still falls far short of the full transformation required by basic sustainability, but it nonetheless constitutes a significant success in some regards. There is hope in this green story, in other words: the partial success of the environmental movement can arguably offer much to learn from as a new generation copes with the challenges of cultural and institutional reorientation required by the technologies of human enhancement.

45. The full text of this remarkable document is available online at http://www.fas.org/oneworld/index.html (accessed November 14, 2008).

7

The Neuroscience of Religious Experience: An Introductory Survey

Stephan Carlson

THE NEUROSCIENCE OF RELIGION pertains to every level of psychological inquiry. Cognitive psychologists like Pascal Boyer ask what distinguishes
the thinking, emotion, and behavior of humans—hence the ability to be
spiritual—from other organisms.[1] In *Religion Explained*, Boyer argues that
religion arises from various cognitive biases of human inference and belief.
Behavioral geneticists like Dean Hamer, author of *The God Gene: How Faith
Is Hardwired into Our Genes*, and Laura B. Koenig and Thomas J. Bouchard
Jr. ask why some people have spiritual experiences and others do not.[2] Finally,
behavioral and social scientists like Richard Dawkins and Daniel Dennett ask
why religion and spirituality persist in the face of technological advances.[3] In
Breaking the Spell: Religion as a Natural Phenomenon, Dennett boldly suggests
that questions about religion and spirituality can finally be tackled by science
because supernatural explanations are no longer necessary.[4]

Thanks to advances in electrophysiology, functional neuroimaging, and
genetics we can now study the neural mechanisms of spirituality and religion.
Dennett contends this is worthwhile because religion affects health, politics,
economics, morality, and even science itself. Richard J. Davidson argues that
studying spiritual practices, especially meditation, may help unravel the neuroscience of our subjective, first-person perspective (consciousness).[5]

In the expansive field of neuroscience, across diverse scientific domains,
at multiple hierarchical levels of complexity, research scientists are exploring
religion in new ways. Neuroscience includes the study of molecules, genes,
proteins, neurotransmitters, receptors, neurons, cognition, emotion, behavior, and complex social interactions. While neuroscientists cannot yet explain

the leaps from genes to religious and spiritual phenotypes, they are pointing us in that direction.

I describe a story of progress across many disciplines converging in contemporary neuroscience. This story culminates in the proposition that we can assess religion as a natural phenomenon. According to this point of view, religion is the product of thoughts, memories, emotions, and behaviors. These processes, in turn, are organized and implemented by the brain. It, in turn, is organized by the action of genes. The theory of evolution, in various iterations, provides the bridging concept that unites these disciplines and that allows neuroscientists to investigate religious beliefs, behaviors, and experiences. Evolutionary theory holds that our brains and bodies—and consequently our *behavior*—are explicable as the product of random variation and natural selection. In fact, R. I. M. Dunbar and Susanne Shultz provide empirical evidence for the social-brain hypothesis, stating that the larger brains in primates evolved because of the demands of processing social information in group living.[6] In genetic parlance, any observable characteristic of an organism, such as its anatomy, physiological properties, or behavior, is called its *phenotype*. As Charles Darwin elaborated in *The Origin of Species* (1859), organisms with particular adaptations that fit a specific environment will be selected to survive.[7] In *The Expression of the Emotions in Man and Animals* (1872), Darwin wrote that "not only has the body been inherited by animal ancestors, but there is continuity in respect to mind between animals and humans."[8]

A Brief History of Neuroscience

Neuroscience began when scholars and physicians performed autopsies and observed injury-induced changes in the nervous system. With the advent of surgery came new insights. Most recently, brain-imaging technologies have produced vastly more subtle and complex pictures of the brain. Contemporary behavioral neuroscientists study the relationships between the structures of the nervous system and behaviors they mediate at the individual level. They perform two types of experiments: In the first, they make *somatic* interventions that alter neuroanatomy (or neurophysiology) and then assess behavioral consequences. In the second, they make *behavioral* interventions and then assess changes in neuroanatomy or neurophysiology. Even more complex is the emerging field of social neuroscience that bridges the gap between neurobiology and social psychology—looking at interactions from animal bonding in prairie voles to human interactions involving trust and violence.[9]

For more than 1,500 years, Western scientists accepted the teachings of prominent Greek physician Claudius Galen (c. 129–199 C.E.) and Andreas Vesalius (1514–1564), the patriarch of anatomical study. Galen and Vesalius said that the fluid-filled spaces of the brain's interior, the ventricles, were responsible for mental processes. Galen taught that the ventricles were filled with *psychic pneuma* (Greek *psychein,* meaning *to breathe,* and *pneumata,* meaning *spirits*), which he described as "the first instrument of the soul."[10] Vesalius, greatly influenced by his predecessor's writings, translated many of Galen's Greek texts into Latin. We see Vesalius's ventricular bias in *De Humani Corporis Fabrica* (On the Workings of the Human Body), where he paid exquisite attention to the ventricles and cursorily drew the cortex.

Ask people in modern society to point to *where* their self is located, and they will almost always point between their eyes. This was not always so, and linguistic vestiges surround us. We "feel from our heart" because Egyptian, Chinese, and Greek thinkers placed the heart at the center of the self.[11] Egyptian priests disposed of the brain through the nose, as it was considered worthless. However, they carefully maintained the heart, which was the seat of intelligence and was necessary in the afterlife.[12] Aristotle (384–322 B.C.E.) claimed that the heart was the seat of the soul and hence cognition.[13] Hippocrates (460–379 B.C.E.) was an exception when he asserted that "men ought to know that from nothing else but thence [from the brain] come joys, delights, laughter, and sports and sorrows, griefs, despondency, and lamentations."[14]

In *Soul Made Flesh: The Discovery of the Brain—and How It Changed the World,* Carl Zimmer locates the transition from cardiocentrism to cerebrocentrism in mid-seventeenth-century Oxford.[15] Even when brain dissection became normative, anatomists, physiologists, and philosophers focused their attention on the ventricles, not on the cortex, which we now know contains a hundred billion nerve cells and a hundred trillion connections. René Descartes (1596–1650) saw the ventricles as the source of the animal spirits that powered the nerves and that produced nonphysical consciousness.[16] He decided that the pineal gland deep in the brain was the seat of the soul because, he said, it collected the spirits from the ventricles. This led Descartes to posit that the mind and brain were composed of different substances: the mind was nonphysical (and immortal), whereas the brain, like the body, was physical and mortal. In one extreme instance, Richard Napier (1559–1634), an English physician, approached physical and psychological problems by utilizing horoscopes, reflecting on the relative contributions of the humors, and invoking angels through prayer.[17]

Zimmer says that Thomas Willis, a fascinating figure who stood at the center of the Oxford Circle, challenged—and refuted—Cartesian dualism.

The Oxford Circle included Christopher Wren, the great architect; Robert Boyle, the chemist who described the laws of pressure; and Robert Hooke, who designed the microscope. Using their diverse discoveries, Willis dissected human corpses with increasing exactness. For example, he discovered what has come to be known as the circle of Willis—which helps ensure neural blood supply—at the base of the brain.

Yet even Willis said that the brain used spirits to control the body and heart. Nevertheless, Willis's new science, neurology, provoked theological debate. Although he was not charged with atheism, he trod carefully through this minefield. Willis published *Cerebri Anatome* (Anatomy of the Brain and Nerve) in 1644, dedicating it to Gilbert Sheldon, the bishop of London. To preempt ecclesial censure, Willis assured the bishop that his book served to "unlock the secret places of Man's Mind and to look into the living and breathing chapels of the Deity."[18] The Oxford Circle eventually evolved into the Royal Society, which catalyzed the scientific revolution that followed.

Franz Joseph Gall and Johann Spurzheim built upon Willis's work. Both Gall (1758–1828) and Spurzheim (1776–1832) promoted phrenology.[19] They also made important neuroanatomical observations, such as describing the functions of white matter as ganglia and grey matter as neurons. They deduced that the brain is folded to conserve space. While much of phrenology is rightfully forgotten, it posited two theoretical insights that dominate neuroscience even today: the first is that the brain is the organ of the mind. To study the human mind—sensation, emotion, cognition, intentional action, and so on—one must study the brain and the nervous system. Their second insight was that different parts of the brain execute different functions.[20]

Following this dictum, Paul Broca, a French surgeon, studied brain-damaged patients. In 1861 he showed that damage to specific brain areas impaired the ability to speak while leaving other aspects of cognition relatively intact. Broca concluded that language could be localized in a particular region of the brain. Carl Wernicke (1874) and Leopold Lichtheim (1885) showed that language was not singular but, rather, subdivided anatomically into speech comprehension, speech production, and conceptual knowledge. Impairment to specific regions creates distinct types of aphasia. Some aphasic patients have poor speech comprehension but good speech production, while others have good speech comprehension but poor production. This suggests that there are at least two speech faculties in the brain and that each can be independently impaired.

Relying on technological innovations in physics to optics, Santiago Ramón y Cajal, a distinguished Spanish anatomist, used microscopes and silver stain, developed by Camillo Golgi, to study brain cells, or *neurons*.[21] In the late 1800s Cajal argued that neurons are the elementary units of the nervous system. Along with Charles Sherrington he then proposed that neurons meet at junctions called *synapses*, where information is communicated from the axons of one neuron to the dendrites of the next. Another important advance was the ionic hypothesis, otherwise known as the Hodgkin-Huxley theory, describing how an electrical current is produced by a neuron, also known as *the action potential*.[22] Focusing on the simple squid axon, Alan Lloyd Hodgkin and Andrew Huxley united the anatomy of the neuron with the physics of electrical signaling and the biochemistry of specific ions—namely, sodium, potassium, and chloride.

Later breakthroughs included identifying receptors and ion channels at the synapse, where neurons communicate with other neurons through chemical signaling. When a presynaptic neuron is depolarized or an electrical current is propagated down the length of the axon, it reaches the axon terminal where chemicals are released into the synapse. These chemicals (neurotransmitters) bind to receptors on the dendrites, or cell body, of the postsynaptic neuron and create a synaptic potential. Further investigation of this chemical communication resulted in the discovery of specific neurotransmitters—such as acetylcholine, glutamate, GABA, glycine, serotonin, dopamine, and norepinephrine—that bind to the next neuron and either excite it or inhibit it to varying degrees. Later researchers discovered that electrical communication occurs between neurons in some instances.

While the neurosciences advanced in the nineteenth and twentieth centuries, psychologists focused on the "black box" of the mind, an entity whose variables cannot be easily measured. Thus, Wilhelm Wundt, the founder of the first psychology laboratory, made methodical observations of the mind and consciousness using introspection.[23] Later pioneers of psychology, like William James and Sigmund Freud, made similar efforts to study consciousness, attention, and personality. However, because they could not get inside the "black box" of the mind it was difficult for them to link their findings to those of neuroscience and related sciences.[24]

In the twentieth century, behaviorists like Ivan Pavlov (who studied the conditioned reflex) and John Watson (who studied fear conditioning) emphasized that mental events are abstract and unobservable by scientific means. Behaviorism became the dominant force in experimental psychology, and behaviorists helped standardize research designs. They also paved the way for the use of animal subjects in brain ablation and genetic knock-out mice techniques to examine behavior.

The limited pertinence of behaviorism to learning, attention, and memory led to the rise of cognitive psychology in the mid-twentieth century. Using the computer metaphor of the mind and related information-processing approaches, cognition was described as a neural network comprised of processing stages. These were often drawn as a series of box-and-arrow diagrams. Championing the neural-network idiom were John von Neumann, who was involved in the development of the first electronic computer, ENIAC, and Alan Turing, who suggested the Turing machine test for computers that could think. They argued that one could understand the cognitive system as a series of steps performed by a computer without reference to brain functioning. Contemporary neuroscientists marry these computational models to neuroanatomy by showing that neurons carry particular information based on input received and transmitted to other neurons.

In the 1970s, Godfrey Hounsfield pioneered the use of macrolevel brain measures to produce computed axial tomography (CT), enabling us to see inside a living body. Angelo Mosso, a nineteenth-century scientist, made an experimental observation fundamental to combining the ideas of CT with dynamic imaging. In *Principles of Psychology* William James says that "[in Mosso's experiments] the subject to be observed lay on a delicately balanced table that could tip downward either at the head or at the foot if the weight of either end were increased. The moment emotional or intellectual activity began in the subject, down went the balance at the head end, in consequence of the redistribution of blood in his system." Mosso recorded the pulsation of blood in the brain during neurosurgical procedures and demonstrated that these cerebral pulsations correlated with mental activity.

Because red blood cells contain iron, they have magnetic properties. Oxygen levels change locally in the brain when the activity of the brain changes. This means that a change in the brain's oxygenation affects the red blood cells, which, in turn, affect magnetic signals generated in an MRI scanner. This provided a way to measure blood oxygen level dependent (BOLD) signals. In functional MRI (fMRI), BOLD signals result from a local change in blood flow that exceeds the change in oxygen consumption.

Using the fMRI technique, we can map changes in brain functioning. In this sense, for the first time in history we can "see" private conscious experience as it occurs in another person's brain.[25] Using the magnetic properties of hydrogen atoms in water, functional magnetic resonance imaging (fMRI) developed in the 1980s. These imaging techniques emerged from interdisciplinary advances in physics, chemistry, math, and computation; each discipline generated new insights that sparked additional insights.[26] From these collaborations came positron emission tomography (PET) and single photon computed tomography (SPECT). The latter measures emissions from radioisotopes to map metabolic and circulatory processes in the body.

Some researchers combined neuroimaging techniques with electrophysiological measures such as electroencephalography (EEG) and magnetoencephalography (MEG). Like fMRI, PET, and SPECT, these techniques are safe and noninvasive. The upside of EEG is that it has good temporal resolution, meaning that changes in an individual are almost instantaneously recorded. The downside of EEG is its low spatial resolution, meaning changes reflect a broad area of the brain. At a macrolevel, these techniques were revolutionary because neuroscientists could now ask questions linking the anatomy of neural circuits with physiological function and behavior.

Cognitive psychologists had already developed experimental designs and information-processing models that could potentially fit well with neuroimaging. It would be no exaggeration to say that the advent of techniques like functional imaging has revolutionized the brain sciences. They may also revolutionize the scientific study of religion.

Back to Darwin: Behavior as Phenotype

Two distinct and competing paradigms explain our capacity for religion. One is supernaturalistic: spirits or souls are the primary actors in religious experience.[27] The other is naturalistic: mechanisms of the brain are the primary shapers of religious experience. The latter paradigm permits us to use data from neuroscience to investigate religious experience. Cartesian mind-body dualists, who assume the mind is nonmaterial, have always opposed the notion that a physical substance like the brain could give rise to first-person subjective feelings, thoughts, memories, and emotions.

In our time Cartesian dualists have yielded to the universal claim that the brain is the part of the body responsible for the functions of the mind. This monism has enormous implications for the study of religion.[28] For example, it has led philosophers and scientists to question free will. If all behavior is the result of physical causes in the material world, is our experience of willing an action more than an illusion?[29] If consciousness is merely biological, there is no afterlife, either, only one's progeny and legacy.

While Darwin did not know about genes, his theory is naturalistic and compatible with contemporary neuroscience. We now know that it is the genotype of organisms that responds to the environment. Over time, the genotype evolves and, in turn, shapes new phenotypes, among them religious and spiritual experience. The most vocal champion of radical Darwinism is Richard Dawkins, who holds the Simonyi Professorship in the Public Understanding of Science at Oxford. Dawkins popularized this genocentric view in his 1976 book, *The Selfish Gene*.[30] In *The Extended Phenotype: The Long Reach of the Gene* (1982), he revised his theory to include the effects that the organism's body and its environment, including other organisms, have on genes.[31] Confronting those who recall the pernicious failures of eugenics advanced in the early part of the twentieth century, Dawkins agrees that absolute genetic determinism is not supported by contemporary science. Studies of inborn errors of metabolism of phenylketonuria, for example, disprove genocentrism.[32]

Consequently, when we speak of a neuroscience of religion, phrases like *a gene for religion* do not have validity. Genes are alterable by the environment, which includes human interventions like psychotherapy and religion. The latter are products of complex gene-environment interactions that occur in distinctive social systems. Thus, we can assign causal roles to both social *and* genetic factors: they all interact in complex, cyclical patterns.

At the same time, contemporary science does support reductionism—the hypothesis that everything in nature is made from a small number of basic constituents that behave in regular ways and are most-readily studied by science.[33] Applying this dictum to the study of spiritual experiences, behavior, and beliefs, we expect them to be chemically, genetically, and neuroanatomically diverse depending on environment and context. At the level of biological pathways, spirituality and religion are not products of single gene-environment interaction. Spirituality and religion comprise thousands of gene-environment interactions that together exhibit diverse phenotypes that we, in turn, recognize as patterns in churches, synagogues, and religious experiences.

As argued by Steven Pinker, humans have various inborn capacities or dispositions.[34] We have long distinguished ourselves from other species by our capacity for religion and spirituality. Nonhumans do not appear to have similarly complex cultural phenomena exhibiting these behaviors. We infer that nonhumans do not have similar spiritual beliefs or experiences either. Our central question is, How we can use neuroscience to explain a cultural phenomenon like religion? A promising answer has come from Dan Sperber, French anthropologist and cognitive scientist. In *Explaining Culture*, Sperber develops a naturalistic approach to culture under the rubric "epidemiology of representations."[35]

In brief, Sperber says that culture is a distribution of beliefs, experiences, and behaviors created among individuals and passed on to other individuals through their minds, brains, and bodies. People wish to share their mental states and to affect the mental states of others. In this way, genes have an impact on cultural evolution via the evolved brains of individuals whose psychological makeup instantiates genes. In other words, biology constrains culture, while culture expresses biology. This is not an exact relationship, because cultures evolve by a parallel process using different forms of transmission. According to Sperber we can explain cultural forms using lower levels of scientific explanation: culture is ultimately a series of gene-environment interactions occurring in a social system passed from one brain to another. This means that spiritual *experiences* might be heritable—products of gene-environment exchange—while religious *behaviors* derive from cultural evolution.[36]

Social neuroscientist Michael J. Meaney impressively illustrates the behavioral transmission of parental traits to their offspring, using his rat experiments to study the relationship between stress and the extent to which a mother licks her pups. In experiments with offspring of low-lick mothers who are raised by high-lick mothers, the offspring of low-lick mothers have similar resiliency as the offspring of high-lick mothers.[37] This behaviorally transmitted resiliency appears to be mediated through alterations in DNA of a glucocorticoid receptor by attachment of a carbon and three hydrogens (methylation). Similarly, our evolved religious behavior and disposition to spirituality are both constrained by evolution and make religion and spirituality possible. In return, genes are affected by religious culture and spiritual experiences. This yields a kind of gene-culture coevolution.[38]

Very Fast, Very Complex Neural Processing

H. Moravec has calculated that real-time human action reflects one hundred billion neurons firing and processing information at one hundred trillion operations per second.[39] This amazing fact highlights the challenge that neuroscientists face when investigating religious and spiritual behavior or any other complex behavior. Francis Crick, nobel laureate and one of the codiscoverers of the DNA molecule, receives credit for this hypothesis. In *The Astonishing Hypothesis: The Scientific Search for the Soul*, Crick posits that "you—your joys and your sorrows, your memories and your ambitions, your sense of personal identity and free will—are in fact no more than the behavior of a vast assembly of nerve cells and their associated molecules."[40]

Crick's statement is explicitly reductionist. Taken out of context it seems to eliminate higher-order scientific explanations.[41] Yet there are multiple levels of scientific explanation, each with its own laws, that are informative and efficient from the perspective of neuroscience. Using explanatory pluralism, neuroscientists approach their subject matter from both higher-order and lower-order levels.[42] Higher-order scientific explanations are useful but are mere approximations filled with exceptions. Lower-order explanations are more reliable but are narrower in focus and not always efficient. Crick's programmatic statement supporting reductionism mirrors other forms of reductionism. Joining him, many contemporary authors argue that higher-order explanations always require the support of lower-order explanations. Higher-order claims must be falsified, corrected, and expanded on in being linked to lower-order scientific explanations.[43]

Some Contemporary Studies in the Neuroscience of Religion

Before proceeding to give an account of the contemporary neuroscience of religion, let's define religion and spirituality. Any such definitions must be provisional, as they pertain to a large number of different behaviors, beliefs, and experiences. One hundred years ago scholars like William James, Edwin Starbuck, and G. Stanley Hall tried and failed to define religion and spirituality, and to this day no clear consensus exists.[44] Some people argue that without formal definitions a field of study loses focus, and its boundaries become too indistinct. This seems plausibly problematic in the humanities, where scholars examine questions of justice, for example, without empirical grounding. In lieu of empirical evidence, differences of opinion sharpen the clarity and rigor of conceptual argumentation.

In the sciences, however, a wealth of evidence suggests that the study of natural phenomena does not require rigorous definition. Because scientists focus on hypotheses, evidence gathering, tests, and repeatability, science is incremental and cumulative. Christof Koch, professor of cognitive and behavioral biology, notes that scientists have made substantial discoveries in the absence of formal definitions.[45]

Dennett defines religions as "social systems whose participants avow belief in a supernatural agent or agents whose approval is to be sought."[46] He notes that this operational definition is not carved in stone. Without suggesting that my definition is better than Dennett's, I will cast my net more broadly. I offer the following definitions: spirituality is the personal experience of the transcendent or something greater than oneself; religion is the specific beliefs, as well as institutional and social aspects, of spirituality.

These definitions help us understand how it is that psychiatrist Robert Cloninger operationally measured what he calls *self-transcendence*. Cloninger says that self-transcendence lies on a continuum in individuals with normal personality traits. Personality traits are the sum of the biological systems unique to each individual that characterize our manner of adapting to life. Cloninger designed an empirically reliable and quantitative scale for measuring personality traits known as a Temperament and Character Inventory (TCI). On this inventory, the self-transcendence trait is made up of (1) self-forgetfulness, or losing one's self entirely in an experience; (2) transpersonality, or connectedness to the larger universe; and (3) mysticism, or the openness to things not literally provable.[47]

Using this scale, Cloninger and others correlate spiritual experience with neuroanatomy, brain metabolism, and cerebral blood flow using neuroimaging techniques. Neural anatomy and function, in turn, link to neurochemistry and genetics.[48] The field of science investigating genetic correlates of stable individual traits is called *the molecular genetics of personality.*[49] By teasing out the diverse biological pathways that lead to similar spiritual behaviors, Cloninger makes far more refined distinctions than those available to phenomenology alone. In genetic terms, Cloninger and his colleagues provide a more systematic explanation of the distinctive phenotypes of religious and spiritual experience.

For example, Jacqueline Borg, Bengt Andrée, Henrik Soderstrom, and Lars Farde explored how the serotonin system might contribute to the biological basis of spiritual experiences. Assuming that drug-induced and naturally occurring spiritual experiences share common neural mechanisms, they examined a particular serotonin-receptor gene variant that is a target for hallucinogens, like LSD. Because some hallucinogens produce spiritual or transcendent experiences or form a component of religious ceremonies, receptors responding to hallucinogens may play a role in nonhallucinogen-induced spirituality. They demonstrated that there is several-fold variability in the number of serotonin receptors across a population and that this difference may explain why people vary in their spiritual interest.[50] On closer scrutiny, these findings are far from robust, and, to my knowledge, the results of this study have not been replicated. Other researchers have evaluated neurotransmitters—such as dopamine, opiate, benzodiazepine, glutamate, and acetylcholine—and their roles in diverse personality traits.[51]

Ulrich Ott and colleagues examined the genetic contributions to individual differences in absorption or self-forgetfulness as described in Cloninger's Temperament and Character Inventory. Absorption is an altered state of consciousness characterized by intensively focused attention.[52] Using Tellegen's Absorption Scale (TAS)—with high levels of absorption forming a phenotype frequently correlated with spiritual experiences, such as prayer and meditation—they assessed the role of the prefrontal cortex in absorbed attention and in schizophrenic symptoms. They hypothesized that absorption is associated with the COMT val158met polymorphism affecting the dopaminergic neurotransmitter system. They found that a group with a particular variant of the serotonin gene had significantly higher absorption scores but failed to find such a correlation with the COMT polymorphism. However, the interaction between serotonin variant and COMT genotypes was significant. This amplifies the functional interaction between the serotonergic and dopaminergic system in potentially influencing spiritual experience. These findings provide neurobiological correlates to what Cloninger calls *self-forgetfulness* and Ulrich et al. call *absorption.*

These spirituality studies are not tied to specific beliefs about god, nor are they tied to traditional practices or distinctive doctrines. Theologians might argue that these studies pertain at best to *perennialism,* the claim that all religious traditions share a common experience—an immediate contact with absolute oneness.[53] Against this theological critique one might argue that disparate traditions have developed techniques independently of one another, all of which lead to similar and measurable outcomes on the subjective-experiential, cellular, and molecular levels.[54] In favor of a general definition of spirituality is that it leads to verifiable hypotheses at the level of the brain dealing with generalizable neuronal correlates. With this operational definition in mind, let us return to the story of neuroscience and its cultural obstacles.

Studying Religious Experiences by Studying the Brain

The idea that the brain mediates conscious and nonconscious religious experiences may not seem controversial. Brain-imaging studies combined with electrophysiological and genetic studies suggest that normal and abnormal temperament (or personality) map onto religious experience with varying degrees of precision.

Daniel Weinberger, a senior investigator at the National Institute of Mental Health, coined the term *imaging genomics*, the use of functional neuroimaging to study genotype-phenotype relationships.[55] In this way, neuroimaging data becomes another measurable phenotype that can be correlated with genetic markers. With this assumption, the capacity for religious experience observed with neuroimaging tools is a phenotypical expression of specific gene influences, of course interacting with environment and experience in shaping the ultimate result.

Researchers studying the neurophysiology of religious experience typically use radioactive tracers (positron-emission tomography, or PET; or single photon–emission computed tomography, or SPECT) or functional magnetic-resonance imaging (fMRI). Physicians treating patients with certain brain disorders—especially complex-partial epilepsy (or temporal-lobe epilepsy), bipolar disorder, and schizophrenia—have long noted that their patients often exhibit striking intensifications of religious experiences.[56] In complex-partial epilepsy, a type of behavioral seizure, high levels of religiosity approaching near total religious preoccupation may occur. In bipolar disorder, a contemporary diagnostic category also known as manic depression, patients in the manic phase may proclaim to be a god, the Son of God, or another deity. Sometimes they report visual or auditory instructions from a deity to carry out a mission. In schizophrenia, religious contents more commonly take the form of persecution by a god.

Of course, believing one's self to be a supernatural being is a delusion, or "a false belief based on incorrect inference about external reality that is firmly sustained despite what almost everybody else believes and despite what constitutes incontrovertible and obvious proof or evidence to the contrary. The belief is not one ordinarily accepted by other members of the person's culture or subculture."[57] Religious experiences evident in schizophrenia, bipolar disorder, and partial-complex epilepsy include hallucinations, "sensory perceptions that have the compelling sense of reality of true perceptions but that occur without external stimulation of the relevant sensory organ."[58]

Neuroscientists use these extreme instances to investigate the neural correlates of religion and spirituality. PET images, for example, help map abnormal circuitry evident in bipolar disorder, schizophrenia, and epilepsy. The cumulated medical history of these conditions was material in establishing a scientific search for the brain regions mediating religious experience.

In one example, B. K. Puri and colleagues use PET imaging to measure the cerebral blood flow in a schizophrenic patient during a state of religious delusion. They then measured the blood flow when the subject was free of delusions.[59] (Puri argued that the medications that treated the patient's psychosis did not affect the second PET scan.) When the patient was delusional, Puri et al. found an increase in left frontal blood flow (frontal lobes modulate speech, working memory, voluntary movement, planning of actions, emotions, and personality) and a mild increase in the left-anterior temporal region (the temporal lobes are involved in processing auditory information and include the limbic system, which is involved in emotion, motivation, and memory processing). Puri et al. suggested that these two areas might be involved with religious beliefs. The increase in left-temporal blood flow during the religious delusional state is congruent with much epilepsy literature. The authors noted that changes in the frontal cortex may be related to attention. This exact case study is one of the lowest levels of empirical evidence and has not been replicated. That said, their findings are coherent with a larger body of evidence.

Many imaging studies focus on meditation, a practice relevant to the neuroscience of religion as a component of religious practice and as a method for achieving spiritual awareness.[60] At its 2005 meeting, the Society for Neuroscience invited the Dalai Lama to give the keynote address, and he spoke about the openness of neuroscientists investigating meditation.[61] Many different types of meditation, including yoga, tantric yoga, yoga nidra, kundalini yoga, and Tibetan yoga, have been studied, making a comparison of the findings difficult. Nancy Andresen argues that one type of meditation practice or one type of experience might differ substantially from another.[62] Are we looking at data about meditative states or traits induced by meditation? A *state* is a transient alteration of experience, such as a change in bodily awareness, relaxation, or emotions. A *trait* refers to lasting change that persists, irrespective of active engagement in the behavior. We also lack comprehensive descriptions of meditation practices. Another methodological issue is that some studies use experienced meditators, while others use novices. Advanced meditators can reliably reproduce their experiences, which is useful when conducting experiments.[63]

For example, Andrew Newberg published a SPECT study of eight Tibetan Buddhist meditators.[64] Their technique involves intense meditation on a visual object until a subjective state of oneness with the image occurs (*self-forgetfulness* or *absorption*). Newberg compared the meditators, both at baseline and during meditation, against data from nine control subjects. Cerebral blood flow in the left-frontal lobe increased during meditation sessions, and Newberg suggests that this relates to enhanced attention. There was a relative decrease in cerebral blood flow in the superior parietal lobule (at the back of the brain). This, he hypothesized, was due to an altered sense of space. Newberg et al. similarly noted an increase in thalamic activity and a thalamic asymmetry in the baseline of the meditators not evident in the controls. The research group argues that these baseline changes are due the effects of long-term meditative practice. They also claim that changes seen in the frontal and parietal cortices during meditation are closely related to the positive mental states described during meditation. These previously discussed studies examined state effects on the brain with imaging techniques.

In addition to understanding how the brain shapes one's spiritual experiences and behavior, meditation also helps us recognize that social environment and interactions shape the brain. A number of studies examine the neuroelectric state effects (EEG changes) during meditation, and there is now general agreement concerning physiological and neurological changes that occur in meditation. Meditation initiates the relaxation response with a decrease in cortisol and blood pressure. There is also some reduction in the galvanic skin response. Subjects who practice meditation over a number of years show distinctive EEG recordings: the most frequently described changes are an increase in frontal theta activity, an increase in temporal theta activity, an increase in central beta spindling, and an increase in generalized coherence. (Coherence is the frequency correlation coefficient, and thus an increase in coherence suggests an increase in similarity in the frequency domain between different brain areas.) All these changes correlate with a level of consciousness that lies between sleeping and waking.[65]

Through interdisciplinary scientific cooperation, these neuroelectric findings have been linked anatomically with particular areas of the brain (right amygdala) that are in turn associated with feeling states, such as relaxation or arousal. These feeling states are linked again to other physiological markers, such as breathing and heart rate, again correlated with the endocrine system (decreasing cortisol involved in stress and increasing beta-endorphins known to reduce pain), and to increases in neurotransmitters in the brain (increasing glutamate, leading to the consolidation of memory and increasing serotonin correlating with positive emotion).

Yet relatively few scientists have looked at Western religious practices. N. P. Azari and his group examined blood-flow changes in six healthy individuals during what Azari termed a *spontaneous religious conversion.*[66] He compared these six against subjects who had had no such experience. He asked his subjects to read a psalm, which invoked intense religious feelings; to read a nursery rhyme, which led to happiness; and to read the telephone directory. A PET scan measuring blood flow was taken during all three states, and the results show identifiable changes that the author claims are specific to religious feeling. Azari says the religious condition showed significantly increased flow in the presupplementary motor area (pre-SMA), in the right lateral frontal cortex, and in the right precuneus. Activity in the pre-SMA supports the necessary preplanning of motor acts and occurs in all behaviors. The precuneus is possibly linked to visual working memory, the lateral frontal cortex with memory retrieval. The religious and control groups differed; however, these changes are not specifically related to experiencing religious feelings. Also, as in any experiment, since only six subjects were involved, replication of this study with a larger subject group would be helpful in confirming the preliminary findings.

Glossolalia, or speaking in tongues, is another Western religious behavior common in Pentecostal religious traditions. It is an involuntary behavior presenting often in several individuals during religious services. The meaning of this phenomenon is interpreted according to specific tradition but is considered to be a connection to the sacred. In 2006, Andrew B. Newberg and colleagues used a SPECT technique to indirectly study brain activity during glossolalia by looking at changes in blood flow to specific brain regions in five women who described themselves as charismatic or Pentecostal Christians.[67] Because glossolalia involves talking or singing, brain activity during glossolalia was compared to brain activity during the singing of religious songs. The investigators found that brain blood flow decreased in the prefrontal cortex, left caudate, and left temporal poles of the brain during glossolalia relative to religious singing. In contrast, brain blood flow increased in the left superior parietal lobe during glossolalia relative to singing. The investigators speculated that blood flow may have been reduced to the prefrontal cortex because blood flow in this brain area increases during voluntary control. Since glossolalia was involuntary compared to the voluntary singing, the investigators reasoned that the prefrontal cortex may have been less active during glossolalia. The investigators had previously associated activity in the superior parietal lobe with a sense of loss of self-identity, and because self-identity seemed to be preserved in glossolalia, the investigators reasoned that this brain area was less active. However, it is unclear how sense of self changed between singing and glossolalia. Temporal lobe–activity changes appeared to be the only changes associated specifically with the glossolalic state.

This brief review of studies identifies patterns of religious experience or religious expression that vary by tradition; we find no universal pattern among brain states that one might term religious. Based on the specific religious experience or behavior we study, we are now able to predict which brain regions are likely unconsciously impacting conscious experience, based on data derived from various neuroscience studies. For example, though central to Buddhist practice, meditation is not the norm in Western religious practice. Glossolalia (speaking in tongues) is common in some Christian evangelical traditions but not in other Protestant faiths or in Catholicism. Hence, we cannot christen a particular brain region the God Spot, nor can we designate which specific brain regions produce religious or spiritual tendencies. The conclusion neuroscientists draw from these data is that humans have the capacity to top-down influence the electrochemical nature of their brains by changing their mental processes.[68] It follows that spiritual experiences, thinking, and ritualistic behavior will have both positive and negative downstream effects on the central nervous system and the body.[69]

The Future Challenge: A Comprehensive Theory

The neuroscience of religion is not as straightforward as many laypersons and scientists would have hoped. We are far from comprehensively understanding religion and spirituality from a neuroscientific perspective, yet it is already grounding our theoretical debates. Empirically it is difficult to verify one mental process or experience as the result of a particular neural correlate, since many functions are assigned to particular brain regions, neurotransmitters, and, hence, genes. This approach considers and accounts for a network of factors beyond our current informational-processing abilities. The best strategy, it seems, is to understand lower-level functions of brain organization—such as neuronal-firing patterns—before we attempt to explain higher-level brain functions—such as spiritual beliefs, behaviors, and experience. Reductionist approaches are also fruitfully complimented by examining how social processes alter the brain and body. This bidirectional programmatic research may eventually meet in the middle. The trend is consistent—integrative explanations are developing that explain how the nervous system is involved in cultural phenomena like religion and spirituality. Whether we can ever explain specific religious beliefs in the brains of religious persons is unclear. Sam Harris—neuroscientist, critic of religious dogma, and author of *The End of Faith* and *Letter to a Christian Nation*—claims to be the first to have characterized belief independent of propositional content at the level of brain activity.[70] Nevertheless, it is clear that neuroscientists will continue to investigate religion and spirituality in all its forms. Verifying claims of neural or genetically inherited biases for spiritual or religious proclivity would have profound implications for theological dialogue. For that reason, this volume on neuroscience and religion is an important part of the debates that must occur.

Notes

1. Pascal Boyer, *Religion Explained: The Evolutionary Origins of Religious Thought* (New York: Basic Books, 2001).

2. Dean Hamer, *The God Gene: How Faith Is Hardwired into Our Genes* (New York: Doubleday, 2004). Laura B. Koenig, Matt McGue, Robert F. Krueger, and Thomas J. Bouchard Jr., "Genetic and Environmental Influences on Religiousness: Findings for Retrospective and Current Religiousness Ratings," *Journal of Personality* 73, no. 2 (April 2005): 471–88. Koenig suggests that studies of adult twins reveal significant heritability of religious experience in the 35 percent to 55 percent range. This is moderately high and matches percentiles observed in mental disorders such as autism, schizophrenia, and attention deficit hyperactivity disorder. These all range between 20 and 50 percent. Similar studies suggest that intelligence, which is very heritable, is 80 percent determined by genes. See also Michael Rutter, "The Interplay of Nature, Nurture, and Developmental Influences: The Challenge Ahead for Mental Health," *Archives of General Psychiatry* 59, no. 11 (2002).

3. Richard Dawkins, *The God Delusion* (Boston: Houghton Mifflin, 2006); Daniel C. Dennett, *Breaking the Spell: Religion as a Natural Phenomenon* (New York: Penguin, 2006).

4. See Dennett, *Breaking the Spell.*

5. Antoine Lutz, John D. Dunne, and Richard J. Davidson, "Meditation and the Neuroscience of Consciousness," in *The Cambridge Handbook of Consciousness*, ed. Philip David Zelazo, Morris Moscovitch, and Evan Thompson (Cambridge: Cambridge University Press, 2007).

6. W. X. Zhou, D. Sornette, R. A. Hill, and R. I. M. Dunbar, "Discrete Hierarchical Organization of Social Group Sizes," *Proceedings of the Royal Society B* 272 (2005): 439–44.

7. Charles Darwin, *On the Origin of Species by Means of Natural Selection, or the Preservation of Favoured Races in the Struggle for Life* (Cambridge, Mass.: Harvard University Press, 1859).

8. Charles Darwin, *The Expression of the Emotions in Man and Animals* (New York: Oxford University Press, 1872).

9. John T. Cacioppo and Gary G. Berntson, eds., *Essays in Social Neuroscience* (Cambridge, Mass.: MIT Press, 2004).

10. Julius Rocca, *Galen on the Brain: Anatomical Knowledge and Physiological Speculation in the Second Century A.D.* (Leiden, The Netherlands: Brill Academic Press, 2003).

11. Stanley Finger, *Origins of Neuroscience: A History of Explorations into Brain Function* (New York: Oxford, 1994), 3.

12. Museum of Science, "Ancient Egypt Science and Technology: Egyptian Afterlife," at http://www.mos.org/quest/mummy.php (accessed November 14, 2008).

13. Charles Gross, "Aristotle on the Brain," *The Neuroscientist* 1 (1995): 245–50.

14. Hippocrates, *The Genuine Works of Hippocrates*, trans. Francis Adams (New York: W. Wood and Company, 1886), 344.

15. Carl Zimmer, *Soul Made Flesh: The Discovery of the Brain—and How It Changed the World* (New York: Free Press, 2004).

16. *The Stanford Encyclopedia of Philosophy*, "Descartes and the Pineal Gland," at http://plato.stanford.edu/entries/pineal-gland/ (accessed November 14, 2008).

17. Carl Zimmer, "A Distant Mirror for the Brain," *Science* 303 (2004): 43–44.

18. William Feindel, "Willis's *Cerebri Anatome*," *Journal of the Royal Society of Medicine* 96, no. 7 (2003): 368.

19. *Britannica Concise Encyclopedia*, "Franz Joseph Gall," at http://concise.britannica .com/ebc/article-9365160/Franz-Joseph-Gall (accessed November 14, 2008).

20. R. M. E Sabbatini, "Phrenology: The History of Cerebral Localization," *Brain & Mind* Magazine (March/May 1997).

21. G. Merico, "Microscopy in Camillo Golgi's Times," *Journal of the History of the Neurosciences* 8 (1999): 113–20.

22. Michael Häusser, "The Hodgkin-Huxley Theory of the Action Potential," *Nature Neuroscience* 3 (2000): 1165.

23. Wilhelm Wundt, "Principles of Physiological Psychology," in *Classics in the History of Psychology*, at http://psychclassics.yorku.ca/Wundt/Physio/ (accessed November 14, 2008).

24. For more on black box theories of the mind, see Volney Gay's chapter "Science, Religion, and Three Shades of Black Boxes" in this volume.

25. P. G. Grossenbacher, "A Phenomenological Introduction to the Cognitive Neuroscience of consciousness," in *Finding Consciousness in the Brain: A Neurocognitive Approach*, ed. P. G. Grossenbacher (Philadelphia: John Benjamins, 2001), 1–19.

26. Giovanni Lucignani and Stefano Bastianello, "Neuroimaging: A Story of Physicians and Basic Scientists," *Functional Neurology* 21, no. 3 (2006): 133–36.

27. R. A. Markus, "Augustine: Man: Body and Soul," in *The Cambridge History of Later Greek and Early Medieval Philosophy*, ed. A. H. Armstrong (Cambridge: Cambridge University Press, 1967).

28. John R. Searle, *Freedom and Neurobiology: Reflections on Free Will, Language, and Political Power* (New York: Columbia University Press, 2007).

29. For example, see the chapter by Jeff Schall in this volume.

30. Richard Dawkins, *The Selfish Gene* (Oxford: Oxford University Press, 1976).

31. Richard Dawkins, *The Extended Phenotype: The Long Reach of the Gene* (Oxford: Oxford University Press, 1999).

32. A. Rosenberg and P. Rosoff, "How Reductionism Refutes Genetic Determinism," *Studies in the History and Philosophy of the Biological and Biomedical Sciences* (July 2005).

33. Ernst Mayr, *What Evolution Is* (New York: Basic Books, 2001), 290.

34. Steven Pinker, *The Blank Slate: The Modern Denial of Human Nature* (New York: Penguin Putnam, 2002).

35. Dan Sperber, *Explaining Culture: A Naturalistic Approach* (Oxford: Blackwell, 1996).

36. A. C. Abrahamson, L. A. Baker, and A. Caspi, "Rebellious Teens? Genetic and Environmental Influences on the Social Attitudes of Adolescents," *Journal of Personality and Social Psychology* 83 (2002): 1392.

37. D. L. Champagne, R. C. Bagot, F. van Hasselt, G. Ramakers, M. J. Meaney, E. R. de Kloet, M. Joëls, and H. Krugers, "Maternal Care and Hippocampal Plasticity: Evidence for Experience-Dependent Structural Plasticity, Altered Synaptic Functioning, and Differential Responsiveness to Glucocorticoids and Stress," *Journal of Neuroscience* 28, no. 23 (2008): 6037–45.

38. C. Lumsden and E. Wilson, *Genes, Mind and Culture: The Coevolutionary Process* (Cambridge, Mass.: Harvard University Press, 1981); L. Cavalli-Sfornza and M. Feldman, *Cultural Transmission and Evolution: A Quantitative Approach* (Princeton, N.J.: Princeton University Press, 1981); R. Boyd and P. Richerson, *Culture and the Evolutionary Process* (Chicago: University of Chicago Press, 1985).

39. H. Moravec, "When Will Computer Hardware Match the Human Brain?" *Journal of Evolution and Technology* 1 (1998): 1–12.

40. Francis Crick, *The Astonishing Hypothesis: The Scientific Search for the Soul* (New York: Charles Scribner's Sons, 1994), 3.

41. C. B. Ogbunugafor, "On Reductionism in Biology: Pillars, Leaps, and the Naïve Behaviorist Scientist," *Yale Journal of Biology and Medicine* 77 (2004): 101–9.

42. Kenneth S. Kendler, "Toward a Philosophical Structure for Psychiatry," *The American Journal of Psychiatry* 162 (2005): 433–40.

43. A. Rosenberg, *Darwinian Reductionism or How to Stop Worrying and Love Molecular Biology* (Chicago: University of Chicago Press, 2006).

44. Brian Zinnbauer and Kenneth Pargament, "Religiousness and Spirituality," in *Handbook of the Psychology of Religion and Spirituality* (New York: Guilford Press, 2005).

45. Christof Koch, *The Quest for Consciousness: A Neurobiological Approach* (Englewood, Colo.: Roberts and Company, 2004), 12. Koch gives the example of attempting to define a gene: "Is it a stable unit of hereditary transmission? Does a gene have to code for a single enzyme? What about structural and regulatory genes? Does a gene correspond to one continuous segment of nucleic acid? What about introns? And wouldn't it make more sense to define a gene as the mature mRNA transcript after all the editing and splicing have taken place? So much is now known about genes that any simple definition is likely to be inadequate. Why should it be any easier to define something as elusive as consciousness?"

46. Dennet, *Breaking the Spell*, 9.

47. C. Robert Cloninger, "A Systematic Method for Clinical Description and Classification of Personality Variants: A Proposal," *Archives of General Psychiatry* 44, no. 6 (1987): 573–88.

48. C. Robert Cloninger, D. M. Svrakic, and T. R. Przybeck, "A Psychobiological Model of Temperament and Character," *Archives of General Psychiatry* 50 (1993): 975–990.

49. R. P. Ebstein, "The Molecular Genetic Architecture of Human Personality: Beyond Self-Report Questionnaires," *Molecular Psychiatry* 11 (2006): 427–45.

50. Jacqueline Borg, Bengt André, Henrik Soderstrom, and Lars Farde, "The Serotonin System and Spiritual Experiences," *American Journal of Psychiatry* 160 (2003): 1965–69.

51. R. P. Ebstein, O. Novick, R. Umansky, B. Priel, Y. Osher, D. Blaine, et al., "Dopamine D4 Receptor (D4DR) Exon III Polymorphism Associated with the Human Personality Trait of Novelty Seeking," *Nature Genetics* 12 (1996): 78–80; J. Benjamin, L. Li, C. Patterson, B. D. Greenberg, D. L. Murphy, and D. H. Hamer, "Population and Familial Association between the D4 Dopamine Receptor Gene and Measures of Novelty Seeking," *Nature Genetics* 12 (1996): 81–84; H. H. Van Tol, C. M. Wu, H. C. Guan, K. Ohara, J. R. Bunzow, O. Civelli, et al., "Multiple Dopamine D4 Receptor Variants in the Human Population," *Nature* 358 (1992): 149–52; H. H. Van Tol, J. R. Bunzow, H. C. Guan, R. K. Sunahara, P. Seeman, H. B. Niznik, et al., "Cloning of the Gene for a Human Dopamine D4 Receptor with High Affinity for the Antipsychotic Clozapine," *Nature* 350 (1991): 610–14.

52. U. Ott, M. Reuter, J. Henning, and D. Vaitl, "Evidence for a Common Biological Basis of the Absorption Trait, Hallucinogen Effects, and Positive Symptoms: Epistasis between 5-HT2a and COMT Polymorphisms," *American Journal of Medicine* 137 (2005): 29–32.

53. The perennialist claim has been strongly challenged on theoretical grounds by many scholars who are characterized as contextualists, such as Steven T. Katz in "Mystical Speech and Mystical Meaning," in *Mysticism and Language*, ed. Steven T. Katz, (New York: Oxford University Press, 1992), 5. For an argument in support of perennialism, see Richard Sharf, "Experience," in *Critical Terms for Religious Studies*, ed. Mark C. Taylor (Chicago: University of Chicago Press, 1998).

54. N. Depraz, J. F. Varela, and P. Vermersch, *On Becoming Aware: A Pragmatics of Experiencing* (Amsterdam: John Benjamins Publishing Company, 2003).

55. A. R. Hariri and D. R. Weinberger, "Imaging Genomics," *British Medical Bulletin* 65 (2003): 259–70.

56. J. Saver and J. Rabin, "The Neural Substrates of Religious Experience," *Journal of Neuropsychiatry and Clinical Neurosciences* 9 (1997): 498–510.

57. *Diagnostic and Statistical Manual of Mental Disorders (DSM-IV-TR)* (Washington, D.C.: American Psychiatric Association APA, 2000).

58. B. K. Puri, P. J. Laking, and I. H. Treasaden, *Textbook of Psychiatry* (Philadelphia: Elsevier Health Services, 2002), 78.

59. B. K. Puri, S. K. Lekh, K. S. Nijran, M. S. Bagary, and A. J. Richardson, "SPECT Neuroimaging in Schizophrenia with Religious Delusions," *International Journal of Psychophysiology* 40, no. 2 (2001): 143–48.

60. B. R. Cahn and J. Polich, "Meditation States and Traits: EEG, ERP, and Neuroimaging Studies," *Psychological Bulletin* 132 (2006): 180.

61. Tenzin Gyatso, the Dalai Lama, "Science at the Crossroads," Mind and Life Institute, at http://www.mindandlife.org/dalai.lama.sfndc.html (accessed November 14, 2008).

62. J. Andresen, "Meditation Meets Behavioural Medicine—The Story of Experimental Research on Meditation," *Journal of Consciousness Studies* 7, nos. 11/12 (2000): 17–73.

63. Cahn and Polich, "Meditation States," 180.

64. Andrew Newberg and J. Iversen.

65. B. E. Jones, "The Neural Basis of Consciousness across the Sleep-Waking Cycle," *Advances in Neurolgy* 77 (1998): 75–94.

66. N. P. Azari, J. Nickel, G. Wunderlich, M. Niedeggen, H. Hefter, L. Tellmann, H. Herzog, P. Stoerig, D. Birnbacher, and R. J. Sietz, "Neurocorrelates of Religious Experience," *European Journal of Neuroscience* 13, no. 8 (2001): 1649–52.

67. Andrew B. Newberg, Nancy A. Wintering, Donna Morgan, and Mark R. Waldman, "The Measurement of Regional Cerebral Blood Flow during Glossolalia," *Psychiatry Research: Neuroimaging* 148 (2006): 67–71.

68. K. S. Blair, B. W. Smith, D. G. Mitchell, J. Morton, M. Vythilingam, L. Pessoa, D. Fridberg, A. Zametkin, D. Sturman, E. E. Nelson, W. C. Drevets, D. S. Pine, A. Martin, and R. J. Blair, "Modulation of Emotion by Cognition and Cognition by Emotion," *Neuroimage* 35, no. 1 (2007): 430–40.

69. M. W. Anastasi and A. B. Newberg, "A Preliminary Study of the Acute Effects of Religious Ritual on Anxiety," *Journal of Alternative and Complimentary Medicine* 14, no. 2 (2008): 163–65.

70. Sam Harris, Sameer A. Sheth, and Mark S. Cohen, "Functional Neuroimaging of Belief, Disbelief, and Uncertainty," *Annals of Neurology* 63 (2008): 141–47.

8

Actions, Reasons, Neurons, and Causes

Jeffrey D. Schall

Imagine: inside, in the nerves, in the head . . . there are sort of little tails,
the little tails of those nerves, and as soon as they begin quivering . . . that
is, you see, I look at something with my eyes, and then they begin quivering, those little tails . . . and when they quiver, then an image appears. . . .
It doesn't appear at once, but an instant, a second, passes. . . . That's why
I see and then think, because of those tails, not at all because I've got a
soul, and that I am some sort of image and likeness. All that is nonsense!
Rakitin explained it all to me yesterday, brother, and it simply bowled me
over. It's magnificent, Alyosha, this science! A new man's arising—that I
understand. . . . And yet I am sorry to lose God!

—Fyodor Dostoevsky, *The Brothers Karamazov*[1]

MOVEMENTS OF INANIMATE OBJECTS such as rocks are explained by external
forces. The occurrence of such forces can be referred to as *causes*. In
contrast, many movements of humans are distinguished from the movements of rocks by *reasons*. These movements produced by humans are described as *actions* directed toward a goal for a purpose and not just as *events*
that happen through a more or less complex chain of causes. Can my typing
these words be explained satisfactorily in terms of a central pattern generator
in my spinal cord that produces muscle contractions in my fingers so that
they bat at the keys on a keyboard? Or must a satisfying explanation include
descriptions of reasons, desires, and plans, such as "I typed this for you to
understand these ideas"?

The central unsolved problem of neuroscience and psychology is to understand how mental entities like *reasons* derive from or at least relate to processes in the brain. But as we come to understand the internal factors of human action in terms of brain function, we must confront the fact that the brain is comprised of neurons and glia that are like rocks in having no interests. To paraphrase Wittgenstein, what, if anything, is left if we subtract brain processes and associated body movements *happening* from an agent *acting?* An answer to this question will require a detailed understanding of how actions arise from brain processes. An answer will also require a deeper insight into the relationship between different levels of explanation that are supposed to translate between the physical and the mental.

To motivate a deeper understanding of correspondence between mental states and brain states, consider the following puzzle: can you think the same thought twice? If every mental state is necessarily grounded in a particular neural state, and if neural states are always changing, then the answer must be no. In fact, if the exemplar-based view of cognition is correct, then perhaps we never do.[2] But, on the other hand, at least in the realm of social interactions it is perfectly intelligible to speak of holding the same thought more than once, consider "New York, New York." Futhermore, the ruminations of a person suffering from obsessive-compulsive disorder are described as repetitive thoughts that preferably would be avoided.[3] On this view, whatever is different between the thought in the first instance and the same thought in the second instance is a difference that does not matter.

A physical example should clarify this point. Consider your signature. You can write it with the pen held normally in your fingers, three or four times. You can write it with a fist wrapped around the pen. You can even write it with the pen held between your teeth or toes. When scaled to the same size, each signature will bear important similarities based on the particular flourishes and shortcuts you have developed. They will be similar enough to one another that, were they on the back of a check, it would be cashed. But upon closer inspection, it is clear that no one of the signatures is a perfect copy of another; they cannot be because of intrinsic variability in the neural and muscular processes that produce the movements.[4] Yet they are all recognized as your signature.

Particular variability in the mapping between mental states and brain states becomes especially important in understanding the difference between voluntary actions and involuntary movements. What did you mean to do? The answer to this question is your intention. The disposition to perform some act is a central feature of an intention, but intention cannot be identified entirely with preparing a movement of the body. A satisfying statement of intention must also answer, Why was that done? Of course, one answer can be the causal path through neurons to muscles, but this seems incomplete in conversation and especially in courtrooms. A satisfactory explanation must address the *reasons* for the action based on preferences, goals, and beliefs. In other words, to judge whether a movement was intended, one must refer to the agent's preferences among the alternative courses of action available, goals that are guiding behavior at the time, and beliefs about which action must be performed under which circumstances in order to bring about the desired object of the intention. A consequence of this is that intentions may not be realized; in other words, success can be judged only with reference to the description of the goal and the conditions under which it could be achieved.[5]

Therefore, defining a body movement as an action depends on context. A purposeful action (a wink) is distinguished from a mere event (a blink) by reference to some intelligible plan, because actions are performed to achieve a goal. In other words, actions have *reasons* (*I did it for . . .*), whereas events merely have *causes* (*It happened because . . .*). Reasons for actions are explanations in terms of purposes, goals, and beliefs. Thus, a particular movement may be intentional under one description but not under another (e.g., a wink or a blink). Furthermore, with human subjects it is possible to distinguish the brain states producing intentional versus unintentional movements, including blinking.[6]

But if all actions are really caused by just neurons firing and muscles contracting, then how can there be any reasons for actions? Reasons are not causes, so what can there be for reasons to do? One answer begins with the evidence for many-to-one mapping of brain states onto movements to elucidate a neural basis of intentional action. If the mapping of neural activity onto movement were one-to-one, then the causal basis of movements would be clear—a particular action follows necessarily from a given brain state as reliably as a reflex. While such an automatic causal process seems an adequate account of certain kinds of movements (e.g., blinks), it cannot provide a satisfactory account of other kinds of movements (e.g., winks). Intended movements are owned (*I did*), while unintended movements are not (*it happened*). In other words, we distinguish the *cause of* from the *reason for* movements.[7]

The goal of this chapter is to articulate how intentional reasons can be reconciled with neural causes through many-to-one mapping of neural activity onto cognition. As noted by several authors, if a given action can arise from different brain states, then the relationship of the behavior to an intention holds in virtue of the content of the representation of the intention and not its neural realization as such—that is, the content that answers, Why did you do that?[8] Thus, a movement can be called an intentional action if and only if it originates from a cognitive state with meaningful content, and this content defines the cognitive state's causal influence. But this analysis depends on whether the brain knows what it means to do. In fact, recent cognitive neuro-science research has described particular brain circuits that register errors and success. Such signals can be used to adjust behavior and provide the basis for distinguishing *I did* from *it happened*, which is just what is needed to feel like we are acting with freedom and responsible power.

Doing

What is an action? Actions are distinguished from mere events by explanations in terms of reasons, goals, and purposes, not just causes. Not too long ago, a well-known natural rock arch in Montana was found collapsed.[9] Originally, it was thought to be vandalism, leading many to ask why someone would do such a thing. Ultimately, though, the arch was recognized to have crumbled from natural causes—erosion. The collapse of the arch just happened. It was not *done.* So it had no reason after all.

What does it mean to do something? One approach to this general question has been to list the order of events involved in the production of a voluntary action.[10] One reasonable list begins with deliberation, leading to a decision, followed by the intention to act, including (somehow) brain processes, nerve conduction, and muscle contraction, followed by a movement of the body. The movement can result in a series of further events extending into the future that appear to be causally related. Which of these is preliminary to action? Which constitutes the action? Which is a consequence of action?

Consider the preliminaries of action. Often one does y by doing x—for example, get grape juice by putting money into a vending machine. We may suppose that x (putting money into a vending machine) is really the action and that y (getting grape juice) is a consequence of the action x. But what if in order to do x one must do w (grasp, position, and release coins)? Then x and y would be consequences of w, and so on. An infinite regress is avoided, though, because some things we do directly, things that are not the conse-quence of some other action. A *basic action* is performed without any prelimi-naries. Basic actions including grasping a coin, moving an arm, releasing the coin, and pressing the button on the vending machine. We cannot say how we grasp, move, and release; we simply will it, and it happens. One possible view, then, is that all that we really do are *basic actions* and that all other oc-currences are more correctly viewed as consequences of these basic actions. The most plausible candidates for basic actions are simple choices or acts of will that are causally first in the sequence of occurrences leading to achieving the desired goal.

Before further defining action, we should distinguish accompaniments of actions from actions themselves. For example, effort is not an action; it only characterizes how some action is performed. Effort alone cannot be done and is not even intelligible apart from the action it characterizes. Our friend Jones can raise her arm directly with no intervening steps, but an intention is always an intention *to do* something, a choice is always a choice *of some action*, and we cannot exert will unless it is the will with which we act.

Another accompaniment of action is a muscle contraction. Simple muscle contractions in most contexts are not basic actions but are necessary conditions for actions, just as brain activation and nerve conduction are commonly inaccessible but always necessary preliminaries to action. Jones neither thinks about nor even has conscious access to the events that make her arm raise and fingers press the dispenser button on the vending machine. However, these preliminary events can, though, under unusual circumstances, become basic actions. Imagine that Jones is an amputee who learns to contract a particular muscle to operate a prosthetic limb. In this case, contraction of this specific muscle per se is the action. Or consider developing prosthetic neural implants.[11] As patients learn to operate the device, producing the correct pattern of neural activation can become a basic action at least until the neural prosthetic operates consistently, at which point the attribution of the basic act may migrate from the production of a pattern of neural impulses to producing the desired movement of the prosthetic. Thus, under typical conditions muscle contractions and neural impulses are necessary preliminaries to actions, but they are most sensibly regarded as events that occur and not as actions that are done.

The definition or attribution of an action depends on context. A purposeful action (pressing one among many buttons on a vending machine) can be distinguished from a mere event (an arm jostled by a friend) by reference to some intelligible plan. However, the distinction of an intended action from an unintended movement can get very involved; it can be difficult to distinguish a wink from a blink.[12] What we correctly call an action, as distinct from a mere event, depends in important ways on what we can easily imagine to interest a person or to form at least part of some intelligible plan. The muscle contraction practiced to operate a prosthetic limb meets this criterion, but the muscle contraction necessary for pushing buttons on vending machines does not. Now, we should not presume that there can be only one correct description that applies to what one does. In fact, alternative descriptions of actions can be equally valid logically.[13] However, people tend to think about an action in just one way,[14] and that identification can change with time relative to the event.[15]

The distinction between body movements that are actions for which an agent can be held accountable and body movements that are events for which the agent is not accountable becomes less clear in various neurological conditions. For example, electrical stimulation of the primary motor cortex in humans produces a body movement, but patients often report that they did not feel as if they willed the movement; instead, they report that the movement was as if someone else moved their body—even though the neural pathway that would normally produce the movement was activated.[16] However, in other instances, patients may confabulate an account of events that incorporates the movement as part of an organized action.[17] In contrast, electrical stimulation of the supplementary motor area can produce the subjective report of the urge to move.[18]

Further neurological evidence for a dissociation of movements of the body from ownership of the movement by the agent is the alien-hand syndrome. This symptom of certain kinds of brain damage involves apparently purposeful movements of the hand that are not owned by the patient. The patient reports that they do not mean for the movement to happen. For example, a patient who underwent a split-brain procedure to treat epilepsy experienced competition between the two hands: "I open the closet door. I know what I want to wear. As I reach for something with my right hand, my left comes up and takes something different. I can't put it down if it's in my left hand."[19]

Wegner has summarized evidence and arguments leading to the conclusion that the sense of willing is not causal but, rather, more correctly understood as a feeling of doing.[20] Fundamentally, Wegner documents conditions in which action and a sense of will are dissociated. Purposeful (seeming) actions can be produced by agents who have no sense of will, no sense of ownership, as in the alien-hand syndrome. On the other hand, agents can report a sense of willing an event over which in fact they have no control whatsoever. Examples include interactions with machines or other agents in which a sense of control over events is attributed incorrectly. An important general conclusion is that the experience of will is entirely distinct from causal connections between brain processes, body movements, and resulting occurrences in the environment. Instead, conscious agency is identified with intentions, beliefs, desires, and plans.

As a result, Wegner has proposed a theory of *apparent mental causation*— conscious will is experienced when agents interpret their own thought as the cause of an action.[21] The more self-conscious, the greater the attribution of authorship of an event. Thinking that it would happen and then experiencing it happening induces a sense that the thinking caused the happening. Accordingly, a sense of will arises from the perception of apparent mental causation.[22] Wegner identifies three sources of the experience of conscious will—*priority, consistency,* and *exclusivity* of a thought about an action. To perceive a thought as the willful source of an action, the thought must occur before the action, be consistent with the action, and not be accompanied by other potential causes. Intentions are thoughts with such properties.

Why would such an illusion evolve? One answer is offered by the creation and analysis of agents in computer-robotic systems.[23] Single movements arising from different causal paths lead to the attribution of different representations or plans. Intelligent agents must include representations of what they mean to do—that is, intentions. Planning ahead, forecasting the future, are also necessary central characteristics of intelligent agents. Internal representations of the world and of the agent in the world are seemingly necessary design features for autonomous agents. The expression of such representations include self-justifying confabulations. The self-representation underlying expressions of intention will take time and experience to develop.[24]

The evolution or construction of an ability to represent plans and reasons for movements is just what allows us to recognize actions in an environment otherwise free of reasons. Of course, with reasons come credit and blame. Certainly in the social arena we are held responsible for the consequences of our actions.

Actions and Consequences

We do actions to achieve goals, satisfy preferences, and produce desirable consequences. What criteria distinguish actions from the consequences that follow? For example, suppose that Jones speaks and, in so doing, answers a question, the consequence of which is that her friend obtains the desired juice drink from the vending machine. One could say that Jones answered a question and that a consequence of her action was that a drink was obtained. However, we could also say without confusion that Jones expressed a desire that had the consequence of satisfying her thirst.[25] Hence, it will be useful to refer to *a drink being obtained* in a way that is neutral with respect to the categories of action and consequences. Thus, we can consider everything that Jones produces when she acts to comprise a temporally extended series of occurrences extending more or less indefinitely into the future. The example we are considering includes the occurrences *words uttered, a request given, a drink obtained, thirst satisfied,* and so on. We must resist the temptation, though, to assume that all elements in this sequence of occurrences are causally and necessarily related to one another.[26]

We can regard *giving a request* as an action and *obtaining a drink* as a consequence. We can also regard *obtaining a drink* as an action and *satisfying a thirst* as a consequence. When a person is performing an extended action—for example, walking to a vending machine to get a drink—it seems more appropriate because of the indirectness with which such a thing is accomplished to call obtaining a drink a *goal* so that when it is achieved it is a *result*, reserving the term *action* for the various immediate means to the goal (for example, walking to a vending machine, grasping a coin, hitting a button).

Because different descriptions of actions can apply, there is no rigid distinction between action and consequence.[27] We are free to construe many occurrences either as consequences or as an extended action. However, even though more than one description of an action and its consequences can be formulated, certain criteria can be applied usefully to classify occurrences as actions or as consequences. Intention is one such criterion. We often say that what a person does is what she intends and that what she does not intend is a consequence, possibly unintended. But this cannot be a sufficient criterion. If it were strictly so, then there would be neither *unintentional actions* nor *intended consequences*, which does not ring true.

When one is engaged in a complex, extended task—for example, earning a college degree—it seems more appropriate to refer to it as a goal, which, when achieved, is a result, and use the term *action* to refer to the various intermediate, more direct, means to the goal—for example, going to class, studying, et cetera. Likewise, it cannot be disputed that unintended consequences occur—for example, coins are dropped into a vending machine, but it is empty. The intelligibility of the concepts of extended goals and unintended consequences may lead us to restrict the term *action* to the occurrence that seems earliest and most causally primary. But this can easily dissolve into restricting the use of action to describe only basic actions of simple body movements. If so, then we are obliged to maintain that no one ever *does* anything except make simple movements of the body. This conclusion does not seem compatible with experience. Alternatively, we can grant that some consequences can be classified either as actions or as consequences of body movements. If we choose this alternative, then we cannot classify occurrences as consequences as opposed to actions solely on the grounds that what precede and produce them can also be classified as actions.

The complexity of the relationship between consequences and actions, the fact that multiple, alternative descriptions can be equally valid, is important in thinking about how rewards guide actions, especially considering that human behavior for rewards can be complex. Discussions of optimizing performance according to reinforcements make critical assumptions about what is being optimized. Is the reward for an undergraduate student in a psychology laboratory the course credit, a modest payment, or the satisfaction of experimental authority?[28]

Actions and Causes

We must be careful in supposing that actions are to consequences as causes are to effects. In the first place, not everything caused by an action is correctly regarded as a consequence of an action.

The effects of one's actions diffuse indefinitely into the future, but we limit the description of the consequences of what one does by some sense of what is important in the situation, what is of interest to us, what is not too distant in time, and what can be explained by referring to intentions. These rather loose criteria are applied to categorize the products of human actions as consequences rather than merely effects of a cause.

We can see that y can be a consequence of x when y is not at all caused by x. Consider the example of Jones raising her hand to ask a question in class versus Jones raising her hand to stretch versus Jones raising her hand to point. The signaling of a question is not caused by Jones raising her hand, because there is nothing more to signaling a question than raising one's arm. In other words, there is no event that is part of signaling a question in class that is not also part of raising one's hand. If x causes y, then two physical (or perhaps mental) events occur in a specific temporal order. But when Jones raises her hand to ask a question, only one physical event occurs. The example of Jones signaling a question by raising her arm contrasts with other sorts of actions, such as Jones interrupting class by shouting a question. In the latter case, there is an obvious causal relation between events that comprise the action (shouting a question) and events that comprise its consequence (interrupting the class).

The distinction between these two examples is captured by distinguishing between doing y *by* doing x and doing y *in* doing x. The second formulation describes cases where doing x is part of what we mean by doing y. Thus, *Jones raised her hand* entails *Jones signaled a question*. Also, *Jones signaled* can be construed as either an action or a consequence of an action. But how are we to distinguish the cases when Jones raised her arm to point or to stretch rather than signal? The framework we are developing hold that which of these three actions Jones actually performs cannot depend on the events she produces. Thus, we must seek other criteria that justify the application of one action description rather than another. Also we must see whether it makes sense to say that these criteria *cause* Jones's action to be of one sort rather than another. In the realm of human and other primate interaction, at least two criteria may be applied to identify an appropriate description of action. The first is the intention with which the action was done. The second is the applicability of social practices or conventions.

Actions like signaling or pointing, or any other socially appropriate actions, are logically dependent on intentions and social practices. The importance of social conventions in defining actions is illustrated potently in an event described by Juarrero, in which a television couple who were actually married portrayed a divorce ceremony and were then judged by religious leaders to be actually divorced.[29] Muslim leaders ruled that according to Islamic law, when the husband pronounced the word *talaq* (meaning *I divorce you*) three times, the wife became totally alienated from her husband and he could not remarry her unless she married another person and that person divorced her on account of marital conflict or she becomes a widow.

Now, given a particular state of affairs, the existence of an applicable social practice logically entails but does not cause a given action. However, the relationship between social practice and the production of an action can be complicated. That is, given an appropriate movement of the body (e.g., raising a hand), it is sometimes sufficient and other times necessary that one intend it or that a social practice applies to it for it to be correct to say that an action takes place. Which of the alternatives—sufficiency or necessity—holds depends on the particular circumstance in which the action is done. For example, suppose Jones raises her arm. Whether she stretches, points, or signals depends on which of these she intends to do—her report of her desired goal. Intending to point, because it is a communicative act, is usually a *necessary* condition for pointing, and if it is performed in the appropriate social context, then it is *sufficient* for pointing as well. However, if the party with whom Jones wishes to communicate is not looking at Jones when she points, then the raising of her hand was not sufficient to achieve her goal of pointing. Thus, the intention to point is a critical part of how the action is interpreted. If this is so, then it seems that there are some actions that, at least in certain circumstances, cannot be done unintentionally. Similarly, other kinds of actions can, in many contexts, be done unintentionally. The professor may misinterpret Jones raising her hand to stretch or point as the action of signaling a question.

Intention

The concept of intention seems necessary to answer questions like, What did she do? and Why did she do that? If a person is asked, they report their reasons, what they intended to do or what consequences were intended. Indeed, people seem compelled to confabulate intentions.[30]

The word *intention* can be used in at least four ways.[31] First, intentions can be expressions of intention, such as *I will buy a grape juice if the vending machine is not empty.* Second, intentions can be ascriptions of intention, such as *Jones intends to buy a grape juice if the vending machine is not empty.* Third, intentions can be descriptions of the intention with which some action is done, such as *Jones's intention of buying a grape juice was to quench her thirst.* Fourth, intentions can be classifications of actions as intentional, such as *Jones bought apple juice instead of grape juice intentionally.* The logical relation between these four uses of the term *intention* is clear. But this means that the word *intention* is ambiguous. The word may refer to a state or an episode or to the intentional object of such a state or episode. In other words, the word can refer to the state of intending or to that which is intended.

Scholars have pondered two basic problems about intentions. First, what sort of state or episode is an intention? Second, how are such states related to actions? More arguments have been developed on these questions than we can deal with here.[32] Nevertheless, few would argue with the claim that at least one of the ideas conveyed by expressions of intention is, other things being equal, a readiness to initiate actions that one believes will obtain the desired goal. The validity of this claim is substantiated by the common tendency for people to profess intentions that they really did not have coupled with the strategy for refuting such a profession. The straightforward refutation of a professed intention involves producing evidence that the claimant made no attempt and exhibited no inclination to do what she said she intended to do. Thus, if Jones is in the presence of a vending machine but does not insert coins and press the grape juice button, then we should conclude one of at least three things: First, Jones changed her mind about obtaining a grape juice. Second, Jones did not realize she was in the presence of the vending machine. Third, Jones did not really intend to obtain a grape juice.

Thus, the disposition to perform some act is a central defining feature of intentions. However, this disposition can be complex. First, a pure intention to perform an act is not intelligible. An additional necessary element of intention is the circumstances under which the act will be performed. Contrast Jones saying *I will get a grape juice* with Jones saying *I will get a grape juice when we arrive at the vending machine if there are any remaining in the machine.* The second statement has a credibility that the first lacks. Accordingly, the intention to do *this* under *these* circumstances can only entail the disposition to do *this* when one believes that *these* circumstances are present. We would have some insight into Jones's intention if the action is simple, such as a movement of her body—*Jones intends to hit the grape juice button on the vending machine.* However, if the action is more complex—*Jones intends to go to the vending machine*—then we still do not know exactly what she will do. She may, for example, take the elevator or the stairs or go to a different vending machine than the one she usually visits. Thus, to understand what someone will do when faced with intentions about complex, extended actions, we must also know her beliefs about the preferred means by which the complex intention can be realized. In other words, what specific actions a person having the intention to do *this* under *these* circumstances will be disposed to perform can only be identified by referring to her preferences and her beliefs concerning what it means to be in *these* circumstances and what actions constitute the doing of *this* or will bring about that *this* is done.

Specifically, because an intention involves the agent's beliefs about the conditions appropriate for its realization and beliefs about the preferred means of its realization, it follows that in addition to any purely behavioral disposition an intention will also involve some conception of what is intended. How a conception of what is intended is involved in a state of intending is complex.[33] One useful approach is recognizing that statements that express or describe intentions possess distinctive formal properties that can be referred to as *marks of intentionality*. These marks belong to all statements about mental phenomena. Two such marks are most revealing about intention. First, a statement of intention may refer to an object or circumstance that may or may not exist. The truth of the statement *Jones intends to purchase a Zippy-Cola* does not depend on the existence of a product called Zippy-Cola. In other words, one can truly intend to carry out some action even if, contrary to one's belief, the action is impossible to accomplish. Second, a statement of intention may contain a descriptive phrase in such a way that one cannot make valid inferences from it. For example, from *Jones intended to purchase a grape juice* and *grape juice is the sweetest drink*, it does not necessarily follow that *Jones intended to purchase the sweetest drink*.

These strictly formal properties of statements of intention are important in at least two ways: The first property shows that intentions by nature may or may not be realized and that a person's intention sometimes cannot be determined by the way she behaves toward the object of the intention (the object on which the agent acts may not be the actual object if the intention—*oops, Jones hit the orange juice button thinking it was the grape juice button*—or the object may not even exist). The second property demonstrates the well-known fact that a particular action may be intentional under one description but not under another. *Jones did not intend to buy the sweetest soft drink. She intended to buy the tangiest soft drink which she believed to be grape juice.* The second property highlights an implication of the first property; the best way to determine a person's intention is to ask her. But, as we saw above, even first-person reports are not always reliable through confabulation or deception.[34]

These points seem to clarify the concept of intention. If intentions may or may not be realized, and if the identity of a person's intention is determined primarily by how she conceives and describes it, then having an intention is mainly a matter of conceiving in a particular way an action or state of affairs while in a state of readiness to do something that will, one believes, achieve or realize the goal directly or indirectly.

We now consider another defining property of intentions. Explanations of actions in terms of intention correspond to or at least allude to the steps of reasoning that the agent could have taken in deciding what to do to realize her intention. Intending to get a grape juice, Jones might say to herself, *I'm thirsty. I would like a grape juice. I need to get enough money before going to the vending machine. I will take the elevator instead of the stairs.* Thus, besides having the disposition described above, an agent who has a certain intention is likely to have thoughts with the logical form of an intention—that is, *I shall do* this *under* these *circumstances.* Of course, the agent does not have to rehearse the thoughts about the intention until the first action is performed. The definition we are establishing only requires that the agent be disposed to conceive such thoughts, for example, when she is deliberating about what to do. So in this sense having an intention is a matter of first being disposed to formulate certain thoughts corresponding to the verbal statements used to express the intention and second to perform actions that one believes will directly or indirectly obtain the intended goal.

Beyond the sense of intention as a disposition discussed so far, we can also conceive of intention in a nondispositional sense—in other words, to engage in acts of intending. Resolving can be regarded as a special case of intending—that is, an immediate consequence of deliberation leading conclusively to choice. Intentions in the form of dispositions may arise effortlessly and unconsciously, like other seemingly spontaneous thoughts, or intentions following deliberation develop explicitly, consciously.

Neural Basis of Agency

Having established a specific sense of intention that depends on an ability to represent plans that constitute reasons for action, we now consider how this can be implemented in the brain. The sense of determinism that seems to trouble those who worry about freedom of will is one in which causal pathways in the brain necessarily produce muscle contractions that, according to this concern, are not free because they were caused. This concern is most clear in the case of reflexes whereby a particular stimulus causes a given body movement. However, humans and other complex animals have brain pathways and processes superimposed on reflex circuits that endow us with very flexible behavior. The thesis I wish to develop is that these higher pathways provide the means for the sense of agency. In short, we do not experience a sense of agency over reflexes because they just happen, but we do experience ownership for nonreflexive, arbitrary body movements executed to accomplish a plan. In fact, the same basic body movement can be owned under some circumstances and not owned under other circumstance.

Consider movements of the eyes. To look around the world, we make rapid shifts of gaze called *saccades*. In fact, we produce two or three saccades every second of the day, amounting to around 150,000 gaze shifts every day. Clearly, we do not experience a sense of ownership around each eye movement, but equally clearly we can. For example, please look right. Did you? Did you feel intentional ownership for that saccade? Thus, movements of the eyes can serve as well as movements of the hands or mouth as a basic action for which to understand the basis of agency. In addition, we will focus on movements of the eyes because research over the last four decades has resulted in a very detailed understanding of the circuits in the brain responsible for the production of saccadic eye movements.

Now, several lines of evidence demonstrate that the same body movement, such as a gaze shift, can originate from distinctly different brain states. For example, a saccade of a given direction and amplitude can be evoked by weak electrical stimulation of a particular site in specific structures in the brain known as the superior colliculus or the frontal eye field.[35] Imagine stimulating a site that elicits a saccade of 10 degrees amplitude directed 45 degrees above horizontal and then stimulating a site that elicits a saccade of 10 degrees amplitude directed 45 degrees below horizontal. Simultaneous stimulation of those two sites elicits a saccade of 10 degrees amplitude directed exactly horizontally.[36] Because the saccade produced when two sites are stimulated is the vector average of the pair of respective saccades, it is referred to as an *averaging saccade*. Electrical stimulation of the brain is not required to produce averaging saccades. We can monitor the eye movements of humans participating in an experiment in which they simply look at spots of light that are presented on a visual display. Imaging a subject who is looking directly at a spot of light when that spot jumps 10 degrees horizontally to the right. After some short period of time necessary for brain processes to locate the target and prepare the eye movement, that subject will produce a saccadic eye movement to fixate the spot. Now, imaging the same subject fixating the same central spot; however, on this trial when the spot disappears from the central location, it is replaced by two spots, each 10 degrees from the center but one located 45 degrees above horizontal and the other located 45 degrees below horizontal. Which of the two spots will the subject look at? In fact, often subjects produce 10 degrees horizontal averaging saccades so that they look at neither spot! A neurophysiological study showed that a given saccade can occur under one circumstance following the activation of a pool of neurons in the superior colliculus responding to a single stimulus and under another circumstance following the activation of two pools of neurons responding to two stimuli presented simultaneously.[37]

This dissociation has been referred to as a *motomere*[38] in parallel with perceptual *metemeres*, which are physically distinct stimuli that evoke indistinguishable perceptual reports.[39] The fact that physically different stimuli cannot be distinguished perceptually means that somewhere between the receptors and the site(s) in the brain where perception is realized information was lost. The existence of motomere equivalence classes has somewhat different implications. First, given that saccades are produced ultimately by a network in the brainstem, the pattern of activation in this network must bear a much more direct correspondence to the saccade that is produced. Consequently, ambiguity inherent in the pattern of activation in the superior colliculus and frontal eye field must be resolved in the brainstem to produce one particular saccade.

Second, evidence for many-to-one mapping of brain states onto movements has important implications for the neural basis of intentional actions. If the mapping of neural activity onto movement were one-to-one, then the causal basis of movements would be clear—a particular action follows necessarily from a given brain state as reliably as a escape response reflex of the Atlantic squid that is mediated by a giant axon ending in a very potent, giant synapse. While such an automatic causal process seems an adequate account of certain kinds of human movements (e.g., blinks), it cannot provide a satisfactory account of other kinds of movements (e.g., winks). Intended movements are owned (*I did*), while unintended movements are not (*it happened*). In other words, we can distinguish the cause of from the reason for movements.[40]

In fact, some have argued that a many-to-one mapping of neural activity onto cognition and behavior provides room for intentional reasons in spite of neural causes.[41] If a given saccade can be the realization of different brain states, then, according to this argument, the dependence of the behavior on an intention holds in virtue of the content of the representation of the intention and not its neural realization as such. The relevant content answers, Why did you do that? Thus, the argument goes, a movement can be called an intentional action if and only if it originates from a cognitive state with meaningful content, and this content defines the cognitive state's causal influence. In other words, the identification of a body movement as purposeful is not derived from the causal chain of events preceding the muscle contraction. It is derived more clearly from the relationship of the body movement to the satisfaction of a goal. According to this argument, freedom to act does not mean freedom from causal pathways but rather freedom to achieve goals.

Indeed, were it not for causal regularity of the world, we could not achieve goals such as satisfying hunger and thirst.

We can apply this argument to saccades. The same saccade can be the outcome of two (or more) distinguishable patterns of neural activity instantiating two (or more) distinguishable representations. The representation of a single focus of activation in the superior colliculus leading to a saccade of a particular vector can be distinguished from the representation of two foci of activation leading to the same saccade through averaging. However, the two mappings of neural representations onto saccades do not have equal status. Averaging saccades are maladaptive, for they direct gaze to neither stimulus; they are errors that must be corrected to achieve the goal of vision. According to this analysis, averaging saccades would be regarded as unintentional errors. If asked, subjects would typically report that they did not intend to shift gaze into the space between two stimuli. In contrast, an accurate saccade to one of the two stimuli would achieve the goal of vision and would more likely be owned as intentional.

It is probably worthwhile to establish an understanding that this principle of a given action arising from different brain states is not restricted to eye movements or even to simple responses to stimuli. For example, a recent functional brain-imaging experiment described patterns of brain activation in subjects playing the ultimatum game in which they must choose to accept or reject a certain amount of money in the context of another player keeping a lesser or greater amount.[42] It was found that choosing to accept a given amount when it was regarded as fair occurred with activation in brain regions associated with reward, while choosing to accept the same amount when it was regarded as unfair occurred with activation in brain regions associated with discomfort.

Thus, according to the argument being developed here, a critical characteristic of intention is sensitivity to consequences. This analysis depends on whether the brain can represent the consequences of actions. Does the brain know what it means to do?

One sign of sensitivity to consequences is adaptation of behavior. In an experimental psychology laboratory such adaptation has been observed when subjects perform repetitive tasks in which they make errors. Commonly, the time to respond to the next stimulus is elevated on trials following errors.[43] More recently, adjustment of response time has been observed under many other conditions. For example, in a task requiring subjects to withhold a planned movement if an infrequent stop signal occurs (known as a *stop signal* or *countermanding task*), response times tend to decrease following a run of trials with no stop signal, and they tend to increase following a run of trials with a stop signal with the increase greatest following trials in which the movement was withheld because of the stop signal.[44]

These adjustments of performance suggest that the brain can register the outcome of the movements it produces. In the early 1990s multiple laboratories discovered a particular signal in brain waves recorded on the scalp of human subjects performing tasks in which they made errors; we will refer to this signal as the *error-related negativity*.[45] A great deal of research has been done and continues to understand how this signal arises and what it does, only some of which will be noted here. One of the first questions asked was, What part of the brain produces this signal? It is generally agreed that the medial frontal lobe including anterior cingulate cortex and the supplementary motor area are the main sources.[46]

Another key question was whether this brain signal was due to the basic action itself or to the consequences of the action. One experiment had a subject produce manual responses to a stimulus and then later informed the subject whether or not the response was correct or in error; a signal just like the error-related negativity was observed, so some now refer to a *feedback-related negativity*.[47] The relationship between error and feedback signals has led to the hypothesis that these brain signals originate from neurons deep in the brain that signal directly whether reward is anticipated or received.[48]

Do these signals arise only if subjects are aware that they have made an error? This challenging question has a complex answer. Under certain circumstances the magnitude of the error-related negativity varies with subjects' sense of accuracy.[49] Under other circumstances the brain signal immediately following an error does not depend on any awareness about the action, but later brain signals do vary according to awareness of the action or consequences and the amount of subsequent adjustment.[50]

Finally, is the medial frontal cortex only active after errors or negative feedback? The answer is clearly no. In fact, the observations of medial frontal activation on correct trials when errors could have been made have led to an alternative account of what this signal is. According to this alternative theory, the medial frontal cortex is actually signaling the amount of *conflict* occurring in the choice process.[51] According to this theory, conflict arises when mutually incompatible response processes are activated simultaneously but cannot both run to completion.

The alternative theories about what the medial frontal cortex monitors and how have been treated as mutually exclusive alternatives, but for the purposes of the argument being developed here, we can appreciate simply that each kind of brain signal—error, feedback about reward, and conflict—exists in the medial frontal cortex and that self-monitoring of performance is accomplished by some complex combination and interaction of these signals. The key point of the argument being developed here is that the neural signals of error, feedback, and conflict not only can be used to adjust behavior but also to provide a mechanistic basis for distinguishing *I did* from *it happened*. In the simplest sense, the hypothesis suggested here is that these monitoring brain signals can distinguish intentional from unintentional movements by their presence; an error-related negativity may follow a wink (did he wink back?) but will never follow a blink. This formulation is consistent with the actor-critic architecture used in robotics.[52] In fact, the feeling of doing described by Wegner may arise from these medial frontal monitoring signals. Certainly, *Oops!* is the archetype of talking to oneself.

As noted above, the error-related negativity can occur after movements that subjects do not appreciate as errors. It must be admitted, though, that at present it is not known whether an error-related negativity will follow a movement that a subject is unaware of producing. This is due to the fact that these brain signals are studied with intentional movements produced for the purpose of some payoff. However, another brain signal arising from the medial frontal cortex has been identified with the production of voluntary movements.

One claim of this argument is that the presence or quality of signals in the medial frontal cortex can identify movements achieving a purpose from those not achieving a purpose or not aimed at any purpose at all. Recall, though, that another attribute of intention (or volition) is the propensity to act. The medial frontal lobe has also been identified with this process through the discovery of another brain signal that precedes voluntary movements by as much as a second (one thousand ms); this is referred to as the *readiness potential*.[53] An extensive and confounding scientific and philosophical literature has developed around this phenomenon, but we will highlight only a few salient observations: First, a widely discussed experiment asked when a sense of volition arose relative to the readiness potential preceding a voluntary movement by asking subjects to notice the time on a special clock at which they first sensed the will to move; the result indicated that the measured time of volition preceded the movement but followed the beginning of the readiness potential.[54] This observation has been replicated and has been suggested by some to provide critical insights into the relationship of brain activation and volition.[55] However, the phenomenon and interpretation have been criticized and extended.[56] One of the most basic problems with the experiment is the assumption that the act of noticing the time does not affect the time that is noticed. In fact, the precise time values depend significantly on how much attention subjects pay to the clock or to their internal state.[57] Furthermore, the act of judging the time of intention itself activates the medial frontal lobe, and the greater the activation, the earlier the judged time.[58] Therefore, the inferences derived from the timing method suggested by Libet cannot be accepted at face value. More important for the argument being developed here is the observation that the readiness potential recorded prior to preplanned, goal-directed movements is of higher magnitude than that recorded prior to spontaneous movements.[59] Also, recall that whereas electrical stimulation of the primary motor cortex in humans produces a body movement that subjects describe as *it happened*, electrical stimulation of the supplementary motor area can produce the subjective report of the urge to move.[60]

Now, as noted above, the concept of intention entails propensity to act, but it also entails the reason for the action in terms of goals and contexts. Other neuroscience research has described how areas in the lateral aspects of the prefrontal cortex contribute to setting goals, establishing plans, and remembering what to do when the time comes.[61] This ability to forecast alternative futures can be identified with the sense of epistemic freedom that is pivotal for arguments for the compatibility of physical determinism and psychological free will. Furthermore, using new techniques to analyze brain-imaging data, it may be possible to infer the intention of an individual performing a simple task consisting of two alternative rules.[62] However, deliberating about alternative futures is difficult when the alternatives are vague and payoffs are uncertain. Such deliberation entails conflict between competing plans. Recall that the medial frontal cortex is recruited when such conflict arises for the purpose of focusing resources on the task at hand. Thus, we can trace at least some of the brain circuits that realize the resolution of such planning conflicts to form an intention to act.

Conclusions

One often hears that things happen for a reason. Of course, if reasons are only in mind of the acting (or interacting) beholder, then we can find reasons anywhere we look as long as we entertain a suitable manner of apprehension. Seeing reasons among causes amounts to an alternative description of events, but humans have always enjoyed a good story.

Notes

1. Fyodor Dostoevsky, *The Brothers Karamazov*, trans. Constance Black Garnett (New York: Modern Library, 1950).

2. L. W. Barsalou, J. Huttenlocher, and K. Lamberts, "Basing Categorization on Individuals and Events," *Cognitive Psychology* 36 (1998): 203–72.

3. G. Doron and M. Kyrios, "Obsessive Compulsive Disorder: A Review of Possible Specific Internal Representations within a Broader Cognitive Theory," *Clinical Psychology Review* 25 (2005): 415–32.

4. P. M. Fitts, "The Information Capacity of the Human Motor System in Controlling the Amplitude of Movement," *Journal of Experimental Psychology: General* 47 (1954): 381–91; D. E. Meyer, A. M. Osman, D. E. Irwin, and S. Yantis, "Modern Mental Chronometry," *Biological Psychology* 26 (1988): 3–67; K. P. Kording and D. M. Wolpert, "Probabilistic Mechanisms in Sensorimotor Control," *Novartis Foundation Symposium* 270 (2006): 191–98.

5. H. Heckhausen and J. Beckmann, "Intentional Action and Action Slips," *Psychological Review* 97 (1990): 36–48; J. Reason, *Human Error* (Cambridge: Cambridge University Press, 1990).

6. B. Libet, "Unconscious Cerebral Initiative and the Role of Conscious Will in Voluntary Action," *The Behavioral and Brain Sciences* 8 (1985): 529–66; I. Keller and H. Heckhausen, "Readiness Potentials preceding Spontaneous Motor Acts: Voluntary vs. Involuntary Control," *Electroencephalography and Clinical Neurophysiology* 76 (1990): 351–61; H. C. Lau, R. D. Rogers, and R. E. Passingham, "On Measuring the Perceived Onsets of Spontaneous Actions," *Journal of Neuroscience* 26 (2006): 7265–71. See also M. Kato and S. Miyauchi, "Functional MRI of Brain Activation Evoked by Intentional Eye Blinking," *Neuroimage* 18 (2003): 749–59.

7. D. Davidson, "Actions, Reasons and Causes," *Journal of Philosophy* 60 (1963): 685–700.

8. J. Kim, "Can Supervenience and 'Non-strict Laws' Save Anomalous Monism?" in *Mental Causation*, ed. J. Heil and A. Mele (Oxford: Oxford University Press, 1995); R. van Gulick, "Who's in Charge Here? And Who's Doing All the Work?" in *Mental Causation*, ed. J. Heil and A. Mele (Oxford: Oxford University Press, 1995); A. Juarrero, *Dynamics in Action: Intentional Behavior as a Complex System* (Cambridge, Mass.: MIT Press, 1999).

9. B. Bohrer, "Forces of Nature May Have Caused Damage to Stone Arch in Montana," Associated Press, Sunday, August 25, 2002.

10. G. E. M. Anscombe, *Intention* (London: Blackwell, 1957); A. Goldman, *A Theory of Human Action* (Englewood Cliffs, N.J.: Prentice Hall, 1970).

11. M. A. Nicolelis and J. K. Chapin, "Controlling Robots with the Mind," *Scientific American* (October 2002).

12. J. L. Austin, "A Plea for Excuses," *Proceedings of the Aristotelian Society* 57 (1956), reprinted in *Philosophical Papers*, ed. J. O. Urmson and G. J. Warnock (Oxford: Oxford University Press, 1961), 175–204; Juarrero, *Dynamics*.

13. Austin, "Plea"; Goldman, *Theory*.

14. R. R. Vallacher and D. M. Wegner, "What Do People Think They're Doing? Action Identification and Human Behavior," *Psychological Review* 94 (1987): 3–15.

15. D. M. Wegner, R. R. Vallacher, G. Macomber, R. Wood, and K. Arps, "The Emergence of Action," in *Journal of Personality and Social Psychology* 46 (1984): 269–79.

16. W. Penfield, *The Mystery of the Mind: A Critical Study of Consciousness and the Human Brain* (Princeton, N.J.: Princeton University Press, 1975).

17. J. M. R. Delgado, *Physical Control of the Mind* (New York: Harper and Row, 1969).

18. Penfield, *Mystery*; I. Fried, A. Katz, G. McCarthy, K. J. Sass, P. Williamson, S. S. Spencer, and D. D. Spencer, "Functional Organization of Human Supplementary Motor Cortex Studied by Electrical Stimulation," *Journal of Neuroscience* 11, no. 11 (November 1991): 3656–66.

19. S. M. Ferguson, M. Rayport, and W. S. Corrie, "Neuropsychiatric Observations on Behavioral Consequences of Corpus Callosum Section for Seizure Control," in *Epilepsy and the Corpus Callosum*, ed. A. G. Reeves (New York: Plenum, 1985).

20. D. M. Wegner, *The Illusion of Conscious Will* (Cambridge, Mass.: MIT Press, 2002).

21. D. M. Wegner and T. Wheatley, "Apparent Mental Causation: Sources of the Experience of Will," *American Psychologist* 54 (1999): 80–91.

22. Wegner, *Illusions*; see also J. W. Brown, "The Nature of Voluntary Action," *Brain and Cognition* 10 (1989): 105–20; S. Harnad, "Consciousness: An Afterthought," *Cognition and Brain Theory* 5 (1982): 29–47; E. J. Langer, "The Illusion of Control," *Journal of Personality and Social Psychology* 32 (1975): 311–28; Libet, "Unconscious Cerebral Initiative"; W. Prinz, "Explaining Voluntary Action: The Role of Mental Content," in *Mindscapes: Philosophy, Science and the Mind*, ed. M. Carrier and P. K. Machamer (Konstanz, Germany: Universitätsverlag, 1997), 153–75; and S. A. Spence, "Free Will in the Light of Neuropsychiatry," *Philosophy, Psychiatry and Psychology* 3 (1996): 75–90.

23. S. J. Russel and P. Norvig, *Artificial Intelligence* (Englewood Cliffs, N.J.: Prentice Hall, 1995); L. Angel, *How to Build a Conscious Machine* (Boulder, Colo.: Westview Press, 1989).

24. For example, see, P. D. Zelazo, J. W. Astington, and D. R. Olson, *Developing Theories of Intention: Social Understanding and Self-Control* (Mahway, N.J.: Erlbaum, 1999).

25. Goldman, *Theory*.

26. S. Vizinczey, *The Rules of Chaos or Why Tomorrow Doesn't Work* (London, Macmillan, (1969).

27. Goldman, *Theory*.

28. C. R. McKenzie, J. T. Wixted, and D. C. Noelle, "Explaining Purportedly Irrational Behavior by Modeling Skepticism in Task Parameters: An Example Examining Confidence in Forcedchoice Tasks," *Journal of Experimental Psychology Learning, Memory, and Cognition* 30 (2004): 947–99; S. Milgram, *Obedience to Authority: An Experimental View* (New York: HarperCollins, 2004).

29. A. Juarrero, *Dynamics in Action: Intentional Behavior as a Complex System* (Cambridge, Mass.: MIT Press, 1999).

30. For example, see Wegner, *Illusions*.

31. B. Aune, "Intention," in *Encyclopaedia of Philosophy*, ed. P. Edwards (New York: Macmillan, 1967).

32. See, for example: G. Ryle, *The Concept of Mind* (London: Hutchinson's University Library, 1949); Anscombe, *Intention*; D. Dennett, *Elbow Room* (Cambridge, Mass.: MIT Press, 1984); and Juarrero, *Dynamics*.

33. Aune, "Intention"; Juarrero, *Dynamics*.

34. Wegner, *Illusions*.

35. Munoz and Schall, "Concurrent Distributed Control."

36. D. A. Robinson, "Eye Movements Evoked by Collicular Stimulation in the Alert Monkey," *Vision Research* 12 (1972): 1795–808.

37. J. A. Edelman and E. L. Keller, "Dependence on Target Configuration of Express Saccade-Related Activity in the Primate Superior Colliculus," *Journal of Neurophysiology* 80, no. 140 (1998): 7–1426.

38. D. L. Sparks, "Conceptual Issues Related to the Role of the Superior Colliculus in the Control of Gaze," *Current Opinion in Neurobiology* 9 (1999): 698–707.

39. See, for example, F. Ratcliff and L. Sirovich, "Equivalence Classes of Visual Stimuli," *Vision Research* 18 (1978): 845–51.

40. Davidson, "Actions."

41. Van Gulick, "Who's in Charge"; F. Dretske, *Explaining Behavior: Reasons in a World of Causes* (Cambridge, Mass.: MIT Press, 1998); Juarerro *Dynamics*.

42. A. G. Sanfey, J. K. Rilling, J. A. Aronson, L. E. Nystrom, and J. D. Cohen, "The Neural Basis of Economic Decision-Making in the Ultimatum Game," *Science* 300 (2003): 1755–58; G. Tabibnia, A. B. Satpute, and M. D. Lieberman, "The Sunny Side of Fairness: Preference for Fairness Activates Reward Circuitry (and Disregarding Unfairness Activates Self-Control Circuitry)," *Psychological Science* 19 (2008): 339–47.

43. P. M. A. Rabbitt, "Errors and Error-Correction in Choice-Response Tasks," *Journal of Experimental Psychology* 71 (1966): 264–72.

44. E. E. Emeric, J. W. Brown, L. Boucher, R. H. S. Carpenter, D. P. Hanes, R. Harris, G. D. Logan, R. N. Mashru, M. Paré, P. Pouget, V. Stuphorn, T. L. Taylor, and J. D. Schall, "Influence of History on Countermanding Performance in Humans and Macaque Monkeys," *Vision Research* 47 (2007): 35–49.

45. M. Falkenstein, J. Hohnsbein, and J. Hoormann, "Effects of Cross-Modal Divided Attention on Late ERP Components: II. Error Processing in Choice Reaction Tasks," *Electroencephalography and Clinical Neurophysiology* 78 (1991): 447–55; W. J. Gehring, B. Goss, M. G. Coles, and D. E. Meyer, "A Neural System for Error Detection and Compensation," *Psychological Science* 4 (1993): 385–90; reviewed by S. F. Taylor, E. R. Stern, and W. J. Gehring, "Neural Systems for Error Monitoring: Recent Findings and Theoretical Perspectives," *Neuroscientist* 13 (2007): 160–72.

46. W. H. R. Miltner, C. H. Braun, and M. G. H. Coles, "Event-Related Brain Potentials Following Incorrect Feedback in a Time-Estimation Task: Evidence for a 'Generic' Neural System for Error Detection," *Journal of Cognitive Neuroscience* 9 (1997): 787–97; H. Garavan, T. J. Ross, J. Kaufman, and E. A. Stein, "A Midline Dissociation between Error-Processing and Response-Conflict Monitoring," *Neuroimage* 20 (2003): 1132–39; E. E. Emeric, J. W. Brown, M. Leslie, P. Pouget, V. Stuphorn, J. D. Schall, "Error-Related Local Field Potentials in the Medial Frontal Cortex of Primates," *Journal of Neurophysiology* 99 (2008): 759–72.

47. Miltner et al., "Event-Related Brain Potentials"; S. F. Taylor, B. Martis, K. D. Fitzgerald, R. C. Welsh, J. L. Abelson, I. Liberzon, J. A. Himle, and W. J. Gehring,

"Medial Frontal Cortex Activity and Loss-Related Responses to Errors," *Journal of Neuroscience* 26 (2006): 4063–70.

48. C. B. Holroyd and M. G. H. Coles, "The Neural Basis of Human Error Processing: Reinforcement Learning, Dopamine, and the Error-Related Negativity," *Psychological Review* 109 (2002): 679–709.

49. M. K. Scheffers and M. G. Coles, "Performance Monitoring in a Confusing World: Error-Related Brain Activity, Judgments of Response Aaccuracy and Types of Errors," *Journal of Experimental Psychology: Human Perception and Performance* 26 (2000): 141–51.

50. See, for example, S. Nieuwenhuis, K. R. Ridderinkhof, J. Blom, G. P. Band, and A. Kok, "Error-Related Brain Potentials Are Differentially Related to Awareness of Response Errors: Evidence from an Antisaccade Task," *Psychophysiology* 38 (2001): 752–60; R. G. O'Connell, P. M. Dockree, M. A. Bellgrove, S. P. Kelly, R. Hester, H. Garavan, I. H. Robertson, and J. J. Foxe, "The Role of Cingulate Cortex in the Detection of Errors with and without Awareness: A High-Density Electrical Mapping Study," *European Journal of Neuroscience* 25 (2007): 2571–79.

51. See, for example, M. M. Botvinick, T. S. Braver, D. M. Barch, C. S. Carter, and J. D. Cohen, "Conflict Monitoring and Cognitive Control," *Psychological Review* 108 (2001): 624–52, and N. Yeung, M. M. Botvinick, and J. D. Cohen, "The Neural Basis of Error Detection: Conflict Monitoring and the Error-Related Negativity," *Psychological Review* 111 (2004): 931–59.

52. See, for example, C. B. Holroyd, N. Yeung, M. G. Coles, and J. D. Cohen, "A Mechanism for Error Detection in Speeded Response Time Tasks," *Journal of Experimental Psychology: General* 134, no. 2 (2005): 163–91.

53. L. Deecke and H. H. Kornhuber, "An Electrical Sign of Participation of the Mesial 'Supplementary' Motor Cortex in Human Voluntary Finger Movements," *Brain Research* 159 (1978): 473–76.

54. Libet, "Unconscious Cerebral Initiative."

55. P. Haggard and M. Eimer, "On the Relation between Brain Potentials and the Awareness of Voluntary Movements," *Experimental Brain Research* 126 (1999): 128–33.

56. See, for example, Libet, "Unconscious Cerebral Initiative."

57. J. A. Trevena and J. Miller, "Cortical Movement Preparation Before and After a Conscious Decision to Move," *Consciousness and Cognition* 11 (2002): 162–90.

58. Lau et al., "On Measuring."

59. Libet, "Unconscious Cerebral Initiative"; Keller and Heckhausen, "Readiness Potentials."

60. Penfield, *Mystery*; Fried et al., "Functional Organization."

61. See, for example, J. Duncan and A. M. Owen, "Common Regions of the Human Frontal Lobe Recruited by Diverse Cognitive Demands," *Trends in Neurosciences* 23 (2000): 475–83; E. K. Miller and J. D. Cohen, "An Integrative Theory of Prefrontal Cortex Function," *Annual Review of Neuroscience* 24 (2001): 167–202.

62. J.-D. Haynes, K. Sakai, G. Rees, S. Gilbert, C. Frith, R. E. Passingham, "Reading Hidden Intentions in the Human Brain," *Current Biology* 17 (2007): 323–28. J. K.

Chapin, K. A. Moxon, R. S. Markowitz, and M. A. Nicolelis, "Real-Time Control of a Robot Arm Using Simultaneously Recorded Neurons in the Motor Cortex," *Nature Neuroscience* 2 (1999): 664–70; Fried et al., "Functional Organization"; D. P. Munoz and J. D. Schall, "Concurrent Distributed Control of Saccades," in *The Oculomotor System: New Approaches for Studying Sensorimotor Integration*, ed. W. C. Hall and A. K. Moschovakis (Boca Raton, Fla.: CRC Press, 2003), 55–82; Nakamura, K. Sakai and O. Hikosaka, "Neuronal Activity in Medical Frontal Cortex during Learning of Sequential Procedures," in *Journal of Neurophysiology* 80 (1998): 2671–87; K. J. D. Schall, "On Building Bridges between Brain and Behavior," *Annual Review of Psychology* 55 (2004): 23–50; M. Tomasello and J. Call, *Primate Cognition* (New York: Oxford University Press, 1997).

9

Human Universals and Human Nature

Thomas A. Gregor

T HE BREAD AND BUTTER of social and cultural anthropology is cultural differences and the different life-ways of distant peoples. Custom and value
vary radically over time and space, and institutions undergo what are, from
our perspective, strange permutations. So it is that in West Africa women may
"marry" women; among the royal families of the Inca, the ancient Hawaiians,
and the ancient Egyptians brother-sister marriage was common; and the Nyar
of India arranged their marriages so that husbands were virtually never the
fathers of their wives' children. What are we to make of such customs? We are,
after all, one species, and as a species we are more genetically homogenous
than most others.[1] Is there a single "human nature," or are we really as different as these practices suggest?

In this chapter I make a case for the importance of universals. After all, if
we seek to link human biology and neural structures to custom, it would seem
reasonable to begin with those customs and behaviors that are general to the
species. By way of illustration, I examine two specific questions: The first, the
cultural recognition of color, focuses on cognition and perception. The second, the origins of the incest taboo, engages fundamental issues of sexuality
and kinship. The two illustrations have a similar intellectual history, in that
they have long been thought to demonstrate the relativity of culture. But with
deeper understanding they are now seen as also reflecting the commonality of
our shared human biology.

Human Nature: Culturally Defined or Universal?

Clifford Geertz has been remarkably influential in cultural anthropology and has consistently argued from the position of relativism: custom must be understood within cultural context. In one of the most widely cited articles on human nature within the discipline, "The Impact of the Concept of Culture on the Concept of Man," he maintains that culture and humans are not separate entities. He says, "If we want to discover what man amounts to, we can only find it in what men are: and what men are, above all other things, is various. . . . The image of a constant human nature independent of time, place, and circumstance . . . may be an illusion. . . . Men unmodified by the customs of particular places do not in fact exist, have never existed, and, most important, could not in the very nature of the case exist."[2]

The view of human nature that emerges from the work of Geertz and his colleagues is that the core of human nature is not what we hold in common but rather the capacity to be socialized in ways that make us different, and capable, for example, of the multiple forms of marriage listed above. At times, anthropologists may be so rejecting of a constant human nature that even the idea of comparison is suspect, in that all cultures are incomparable apples and oranges. Mark Hobart, for one, labels comparison as intellectual snake oil purveyed by charlatans. He illustrates his point by examining a human experience that should be universal but which he claims is not: that of eating. "Everywhere," he says, "animals and people eat. Is this not a universal that underwrites translation?"[3] Following Hobart, it does not. Hence, in Bali there are eight words for eating, two of which describe consumption by high priests and Brahmins or are used when inferiors address princes. Another of the terms describes eating by lower castes and some animals, while still additional words reflect degrees of politeness and formality. Hobart says that these distinctions are so fundamental that they do not have a common core. Rather the experience of eating for the Balinese is so intertwined with social distinctions and the nature of being a Balinese human that it can not be understood apart from it.

Hobart has a point, in that culture surely infuses human nature. But is there not a universal experience of hunger, chewing, swallowing, taste, and satisfaction that unites us as a species? It is likely that Hobart's response would be very much like that of Geertz, who reflects a still-influential relativistic perspective. For Geertz, cultural universals are generally trivial. Far from expressing what is essential about humans (which is their capacity for diversity), they are either fake or tautological: "That everywhere people mate and produce children, have some sense of mine and thine, and protect themselves in one fashion or another from rain or sun are neither false, nor, from some points of view, unimportant, but they are hardly very much help in drawing a portrait of man that will be a true and honest likeness."[4]

Geertz's point is well taken. When we look at efforts to pinpoint what we have in common, the result is often a trait list, and sometimes a rather dull one. Historically, such projects began in the early twentieth century with Wissler's "Universal Cultural Pattern," Kroeber's "Total Cultural Pattern," George Peter Murdock's *Outline of Cultural Materials*, and Clyde Kluckhon's "Universal Institutional Types."

One of the more recent and comprehensive efforts to put together such a list of common institutions and traits is that of Donald Brown, who asks, "What do *all* people in *all* societies, *all* cultures, and *all* languages have in common?"[5] The answer is an extended list, one which continues to grow as more is learned about the world's cultures. They describe what Brown calls the universal people, or UP, whose "ethnography is a description of every people, or people in general."[6] In the list below I present the UP as alphabetically arranged and abbreviated from Brown's research.[7]

Excerpted and Edited from
Donald E. Brown's List of Human Universals

1. anticipation
2. attachment
3. beliefs about disease
4. binary cognitive distinctions
5. classification
6. classification of age, body parts, colors
7. dominance/submission
8. fairness (equity), concept of
9. fear of death
10. generosity admired
11. gift giving
12. good and bad distinguished
13. gossip
14. government
15. grammar
16. group living
17. hairstyles
18. healing the sick (or attempting to)
19. hope
20. hospitality

21. hygienic care
22. identity, collective
23. incest between mother and son unthinkable or tabooed
24. inheritance rules
25. insulting
26. judging others
27. language employed to manipulate
28. language not a simple reflection of reality
29. law (rights and obligations)
30. leaders
31. linguistic redundancy
32. logical notion of *and, equivalent, general/particular*
33. logical notion of *not*
34. logical notion of *part/whole*
35. magic to sustain life, to win love
36. male and female and adult and child seen as having different natures
37. males engage in more coalitional violence, dominate public/political realm
38. males more aggressive, prone to lethal violence, more prone to theft, travel more
39. marriage, division of labor by sex and age, females do more direct childcare
40. stop/nonstop contrasts (in speech sounds)
41. medicine
42. melody
43. mental maps
44. metaphor and metonym
45. morphemes
46. mourning
47. murder proscribed
48. music, children's music, relationship to dance, ritual
49. musical variation
50. overestimating
51. objectivity of thought
52. person, concept of personal names
53. phonemes
54. phonemes defined by sets of minimally contrasting features
55. phonemes, range from ten to seventy in number
56. planning for future
57. play, play to perfect skills

58. poetry/rhetoric
59. poetic line, uniform length
60. poetic lines, repetition and variation
61. poetic lines demarcated by pauses
62. polysemy (one word with related meanings)
63. proverbs, sayings
64. resistance to abuse of power
65. risk-taking
66. self-image, awareness of
67. self-image, manipulation of
68. self-image, wanted to be positive
69. sex differences in spatial cognition
70. shame
71. statuses, ascribed and achieved
72. statuses distinguished from individuals
73. statuses on other than sex, age, or kinship bases
74. stinginess, disapproved
75. succession
76. sweets preferred
77. symbolism
78. verbs
79. violence, some forms proscribed
80. vocalic/nonvocalic phonemic contrasts
81. tabooed foods and utterances
82. taxonomy
83. territoriality
84. thumb sucking, tickling
85. time, cyclicity of
86. weapons

Some of the items on the list appear to be inevitable, even tautological in the sense of Geertz's complaint, in that it is hardly surprising that human beings universally feel pain, disapprove of stinginess, and must cook their food. Geertz found such lists to be not only trivial but also misleading. "Is the fact that marriage is universal," he writes, "as penetrating a comment on what we are as the facts concerning Himalayan polyandry, or those fantastic Australian marriage rules, or the elaborate bride-price systems of Bantu Africa?"[8]

These criticisms are off the mark, in that clearly many of the universals from the list above are inherently significant, including such items as binary cognitive distinctions, cyclical concepts of times, tabooed speech, violence, sex differences in spatial reasoning, contrasts of vowels and consonants, and specific logical devices (see especially items thirty-two to thirty-four). Further, if we focus exclusively on gender, we find significant commonalities that reflect on masculinity:

> male and female and adult and child seen as having different natures
> males dominate public/political realm
> division of labor by sex and age
> males engage in more coalitional violence
> males more aggressive, prone to lethal violence, and more prone to theft
> males, on average, travel greater distances over lifetime
> females do more direct childcare

To Brown's list we may add an entire array of cognitive traits shared by all the universal people, including "an intuitive physics with a core concept of an *object*, which, in motion, has its own internal momentum; a sense of engineering by which we understand tools as objects made with a purpose; an intuitive psychology by which we impute motives to others and see them as containing a soul-like essence; and a sense of probability."[9] Were our senses and reasoning accurately attending an external reality around us, these commonalities would be less interesting. Instead they appear to be rough and ready rules of thumb, which are sometimes erroneous. Hence we succumb to and are routinely misled by the same illusions, logical lapses, and skewed sense of probabilities. What we learn from these uniformities is that there is a panhuman way of perceiving and reacting to the world, one based on similarities of brain structure and perception.

That relativistic approaches to culture and the human condition flourish in anthropology is in part a disciplinary preference for metaphorical trees rather than forest. Geertz, as with many of his colleagues, is tree-focused. Hence, he sees anthropologists as those who must "hawk the anomalous" and "peddle the strange." They should be "merchants of astonishment."[10] Brown's objective is quite different. Since we are ourselves one of the UP tribes, there are few surprises on the list, and the enterprise does not meet Geertz's astonishment test. For those who will step back for a view of the forest, however, the universal people speak poignantly with what Pinker calls the "voice of the species." In its "sheer richness and detail," he writes, we see that custom transcends cultural and historical differences and bespeaks a fundamental biological and social unity of human kind.[11]

Let us return to the case of marriage to demonstrate that the search for universals results neither in tautology nor triviality. Despite Geertz's objection and the strange (to us) examples at the outset of this essay, marriage *is* a universal, or, at the very least, a near-universal human institution. In its most common form, a marriage binds men and women in an economic relationship based on the sexual division of labor, establishes a household, permits sexual access, and legitimizes children as members of their kin group.[12] To be sure, there are significant differences in marriage, in that many societies practice polygyny (polyandry is very rare and limited to societies where the labor of more than one man is needed to support a family), and there are rare instances where households are formed on the basis of sibling relationships (spouses may actually live separately).

The case of woman-woman marriage in West Africa cited above is less of a challenge to a universal view of marriage than it might seem, since it has no sexual implications. A well-to-do but childless older woman can take a younger wife, who will bear a child that will belong to the older woman's kin group. True same sex marriage is known primarily in Western societies, where it may be, as in the United States, contested.

Some of the apparent exceptions to marriage may be understood with the concept of *focality*. That is, societies that recognize alternate marriages may not accord them the same status as monogamous or polygamous marriages between the sexes. Hence in our own society the focal concept of marriage refers to a relationship between a man and a woman. This may be in the process of change, as there is increasing acceptance of homosexual marriages. The Catholic Church already institutionalizes the marriage of a nun to Christ. But it would be difficult to argue that these marriages are all culturally the same.

The near-universal dimensions of marriage reflect the evolution of human reproduction and consequent male-female bonding. In brief, our near relatives, monkeys and apes, do not walk upright. Our own transition to upright stature, which is acknowledged as the crucial step toward our human status, required far-reaching skeletal changes to the hips and pelvis, including the relative narrowing of the birth canal. Ultimately natural selection favored increasing intelligence and a far larger cranial capacity. How could females give birth to increasingly large-headed offspring? The answer was to give birth to small-headed and increasingly immature offspring. A human infant is born with only a quarter of its adult brain weight. By contrast, a rhesus macaque is born with about four-fifths of its adult brain weight. As a consequence, human offspring are born immature and dependent on mothers who must nurse them. Males who will provision and protect the mother-infant pair are a part of the equation. Marriage, the sexual division of labor, and the relationship of parents to dependent children, may thereby have roots in early evolutionary experience. While acknowledging Geertz's interest in the cultural differences in marriage practices, we can also stand back and learn something of the human condition by observing what is general about it.

We now offer two examples of such a universalizing perspective. The first is the cultural recognition of color, which sheds light on panhuman processes of cognition and perception. The second example is that of the origins of the incest taboo, which touches on one of the oldest problems in anthropology and illuminates fundamental issues of sexuality and kinship. The two illustrations have a similar intellectual history, in that for a period they were thought to demonstrate the diversity of culture and, in the case of color, a gold standard of cultural relativism. But with deeper understanding, we see them as also reflecting the commonality of our shared human biology.

Universals and Culture: The Case of Color

The role of relativism and universals in culture came into sharp focus in 1969 with the publication of *Basic Color Terms* Berndt Berlin and Paul Kay on the categorization of color. The perception of color is determined by the frequency of the light stimulus, its purity (the light we perceive, other than laser beams is almost always of multiple frequencies), and the strength of the signal (brightness). So sensitive is the eye that by varying frequency, purity, and brightness under laboratory conditions we are capable of making thousands of distinctions and, conceivably, recognizing as many colors. It would appear that nature is virtually seamless in presenting us with a vast spectrum of colors. Languages, however, pick and choose and do not mark more than eleven basic colors, a basic color being defined by its frequency of use and its irreducible semantic reference to color. (Hence blue counts, but puce and robin's egg blue do not, the first due to its scarcity, the second due to its compound nature.) In English, we recognize all eleven colors, specifically red, yellow, green, blue, brown, purple, pink, orange, black, white, and grey.

One of the first field investigations of color categories was led by W. H. R. Rivers's expedition to the Torres Islands in the Pacific in 1898. He found color-term systems quite different from our own, including languages that merged the eleven distinctions we make under fewer terms. Subsequent investigators have found languages that have only three basic color terms— white, black, and red. In such systems, red included many of the warm colors we differentiate, such as orange, yellow, brown, pink, and purple. There are a few languages that have but two color terms—black and white—which we might best translate as relative lightness and darkness. The varying definitions of color terms was forceful evidence in favor of the diversity of culture, the cultural definition of experience, and the power of language in defining reality: "There is no such a thing as a natural division of the spectrum. Each culture has taken the spectral continuum and has divided it up on a basis that is quite arbitrary."[13]

Systematic experiments with speakers of different language at first confirmed the arbitrariness of culture in dissecting and labeling the spectrum. The data demonstrated that the boundaries of color categories did indeed vary across cultural boundaries. For example, among the Mehinaku of Amazonia, blue and green are referred to by one term. The apparently arbitrary way in which culture dissects the spectrum seemed analogous to language. Spoken words, with a few onomatopoetic exceptions, give no clue as to their meaning unless one knows the language. The case of color similarly demonstrated the arbitrary cultural construction of the most basic level of perceptions and senses. Or so it seemed.

From the start, there were reasons to suspect this conclusion. The biology of human color perception is everywhere the same, utilizing retinal rods and cones that are sensitive to red, blue, and green light. Further, no one demonstrably claimed that the absence of a color term in a particular language meant that its speakers could not perceive that hue. For example, the Mehinaku normally make no distinction between blue and green. But when they refer to a specific green item that must be distinguished from a blue one, they will say "the lizard one." More tellingly, as Berlin and Kay noted in their pathbreaking study, despite differences in the spectral boundaries of color terms they seemed too "translatable." The best example of a red in one language looked very much like a red in another, even though one language might include orange, yellow, and pink under the larger umbrella of red.

I had my own lesson in the translatability of colors in the course of my own field research among the Mehinaku. I had left the field for a month's vacation, and I promised the women of the village that I would bring back dress material. They had asked for material that was *really* red. "But what," I asked myself in the dry-goods shop, "is *really* red for the Mehinaku?" Having no other standard than my own, but with Berlin and Kay's theory under my belt, I took the plunge and ordered forty yards of fire-engine red cotton cloth for the women. The village women were entranced. One of them approached me and said, "That is *really* red."

This is more than a shaggy-dog story, in that other colors in other languages generally fit this model. *Boundaries* may vary, but in all probability it is the universal structure of the human eye and its underlying neurological processes that make the experience of seeing colors so similar from society to society. What began as a project in demonstrating the relativity of culture and its arbitrary coding of sensory experience ended in the discovery of human universals, or at least near universals. Marshall Sahlins, an anthropologist who normally emphasizes the diversity of culture, has identified this finding as "among the most remarkable discoveries of anthropological science."[14]

Universals and Sexuality: The Case of the Incest Taboo

One of the oldest questions in anthropology is the origin and ubiquity of the incest taboo, a prohibition on sexual relations between individuals who believe they are too closely related. Often the taboo is extended to individuals who are objectively nonrelatives. In traditional China, persons with the same surname could not get married even though they were unrelated. These differences in the scope of the taboo encouraged anthropologists to assume the taboo was relative to the society in which it appeared. Hence, Rodney Needham could write in 1971, in the midst of a wave of intellectual relativism in anthropology, that the incest taboo is "not a universal. . . . There are as many different kinds of incest prohibitions as there are discernable social systems."[15]

Needham's insistence on culture as the source of the taboo resonates with the interpretation of varying color terminology. The most salient aspect of culture in forming the taboo from Needham's and others' perspectives is the system of marriage.[16] Above all, the taboo creates a contrast between blood kin, who are unmarriageable, and in-laws, with whom one may cooperate and form alliances. The value of such alliances was such that one of the founders of modern anthropology, E. B. Tylor, saw it as a necessity, in that early humans could either "marry out or die out." In fact, followers of taboos will sometimes offer such explanations, but their narratives may have an after-the-fact tone that does not leave us better informed. Here is the Mehinaku chief's explanation of the taboo on sibling incest, as I recorded in on one of my early field trips:

Q: Why not marry your sister?

A: No, no. You do not have sex with your sister. Your sister marries a man who becomes your brother-in-law, and you go fishing with him. Who would you go fishing with if you married your sister?

This explanation makes sense, in that a violation of the incest taboo would obliterate the distinction of in-laws and blood relatives and that the Mehinaku labor system is organized with cooperation between in-laws.

But the incest taboo may be grounded on more than varying social structure:

Q: Suppose you had lots of people to go fishing with. Could you marry your sister?

A: No; a sister is too close to marry. Sex would be wrong and stupid with a kins-woman; it would be like an animal.

Q: Why is it wrong?

A: Do *you* have sex with your sister? Do Americans think incest is good?

And here we hit bottom. Incest is deeply wrong from the perspective of this informant, but he is not able to fully explain it.

The cultural explanation of the incest taboo as reflecting marriage practices suggests it is variable. But as in the example above, if we turn our attention to the more intimate circle of kin, the taboo seems less exotic. It is more frequently enforced, more powerful, and very similar from culture to culture. The line that is drawn most dramatically is between marriages between first cousins and those within the nuclear family. Marriage between cousins is actually the preferred marriage in many small-scale societies, in that it links intermarrying families with strong bonds of alliance through both blood kinship and in-law relationships. On the other hand, brother-sister and father-daughter relationships are all but universally tabooed (see below), and mother-son incest is never permissible.

The taboo thus intersects with issues of sexuality, reproduction, family structure, kinship systems, and culture. It is therefore bound to be overdetermined in terms of how it came to be and how it is sustained, and as such there are multiple theories as to its origins and functions. Our attention is drawn to those determinants that seem to operate through the universals of human biology, the most compelling of which is the effect of inbreeding. As is well known, deleterious but recessive genes are more likely to be expressed when parents are close biological relatives. In nonhumans the negative impact is undeniable, and the percentage of nonviable offspring increases with each generation of successive incestuous matings. For this reason, incest is rare within the animal kingdom, especially among primates and mammals, but is also unusual among birds, reptiles, amphibians, and insects. Measuring the impact among human is not easy, since incestuous relationships are often hidden and probably occur in less than 1 percent of the population. They are often associated with extreme psychopathology within the nuclear family with congenital defects, with drug use, or with illness, all of which confound cause and effect in measuring the viability of offspring.[17] Nonetheless, a recent metastudy of incestuous relationships—as well as extrapolations from studies of consanguineous marriage in non-Western societies (for example, in Southern India cousin marriage and uncle-niece relationships are *preferred* relationships)—allow us to make reasonable estimates. The offspring of siblings or parent-child unions suffer death or severe disability at a rate of between 22 and 36 percent.[18]

Granted that inbreeding is dangerous, but what are the proximate mechanisms by which this hazard translates into avoiding incest and creating a powerful taboo? One answer is that it may simply be a matter of observation of the disastrous consequences of inbreeding.[19] From recognition to aversion and taboo would be a relatively short step, in which the act is associated, for example, with vengeance of spirits and gods. Durham finds some supporting evidence in a small cross-cultural study, in which more than half the societies in his sample avoided incest because they feared they would otherwise have "sickly, weak, and half-witted children."[20]

The American folk theory of incest is similar to many of the cultures in Durham's sample, in that we recognize the dangers of inbreeding. But for us, as for perhaps all the world's peoples, the taboo is laden with emotion that goes well beyond a rational assessment of the negative consequences of inbreeding. Hence we avoid many dangerous practices (or even willingly engage in them) without the assistance of a taboo. For example, smoking has health outcomes that are known and appreciated, but the aversion to the habit comes nowhere near the emotional intensity of incest avoidance. We are reminded that a taboo is not simply a rule. Rather it is a prohibition that is laden with anxiety and religious intensity that Freud aptly termed *holy dread*. The word *incest* itself has roots that signify unclean, impure, and repugnant.

To underscore the irrational component of the taboo, we can do no better than cite the work of Jonathan Haidt, who devised a cleverly subversive technique for exploring the nature of our moral reasoning with respect to incest. Consider the following story, which was presented to experimental subjects:

> Julie and Mark are brother and sister. They are traveling together in France on a summer vacation from college. One night they are staying alone at a cabin near the beach. They decide that it would be interesting and fun if they tried making love. At the very least it would be a new experience for each of them. Julie was already taking birth-control pills, but Mark uses a condom too, just to be safe. They both enjoy making love, but they decide not to do it again. They keep the night as a special secret, which makes them feel even closer to each other. What do you think about that? Was it okay for them to make love?[21]

Haidt discovered that those who hear the story are vehement in their conviction that Julie and Mark did something very wrong, but they were "morally dumbfounded" when they tried to explain just what it was. Was it that they would be psychologically disturbed if they had sex? The story tells us they were not and that it actually enhanced their future relationship. Was it the effect on others? No, we are told that the relationship was kept secret. Was it the danger of inbreeding? Well, Julie and Mark used two forms of contraception. Although we can martial many arguments as to why incest is wrong, it is wrong at a level that is not easy to articulate.[22] The aversion lies so deep within us that we can not easily expose it to light of day.

The Westermarck Effect

What is the proximate source of the aversion? In 1895 Eduard Westermarck provided us with an answer that focused our attention on the early years of childhood. "Generally speaking," he wrote, "there is a remarkable absence of erotic feelings between persons living very closely together from childhood. Nay more, in this, as in many other cases, sexual indifference is combined with the positive feeling of aversion when the act is thought of."[23] The history of the Westermarck effect parallels that of color categories. It was initially disparaged as anthropologists sought explanations in culture for the taboo. But in recent years the evidence, which has focused primarily on sibling relationships, has become persuasive and is now widely accepted.

Two empirical studies are particularly impressive. In the first, Arthur Wolf (1995) examined two forms of arranged marriage, which until the middle of the twentieth century were followed in both China and Korea. In the first, bride and groom did not meet until the day of their wedding. In the second, which Wolf refers to as *minor marriage* (referring to its secondary status rather than marriage between minors), the groom's family "adopted" a girl and raised her in intimate association with their son to eventually become their daughter-in-law (she was referred to as *sim-pua*, or *little daughter-in-law*). The children were raised by the same adults, and they were "in contact almost every hour of every day. Until seven or eight years of age they slept on the same *tatami* platform with [the boy's] parents; they eat together and play together; they are bathed with the other children of the family in the same tub. . . . They are free to behave as though they were siblings until they are designated husband and wife."[24] Ultimately, the children grow to maturity and marry.

In the *sim-pua* marriage Wolf had a natural experiment. Would the relationships of *sim-pua* brides and their husbands reflect the Westermarck effect? Wolf found evidence that they did. When *sim-pua* brides were adopted at an early age, their fertility was 40 percent lower than women in ordinary marriages, divorce was 300 percent higher, and the women in these marriages were twice as likely to engage in extramarital sex than were women who married a stranger, and, in one community study, their husbands were three times more likely to frequent prostitutes than men who had been married to women they did not know.[25]

Absent the Westermarck effect, we might well have expected the *sim-pua* marriage to have a better chance of success than the marriage of strangers, in that both partners would have had years to learn how to get along together. But the core of the problem appears to be sexual aversion. A couple raised together are like siblings. They find the prospect of sexual relations "embarrassing and uninteresting," and their problematic relationships are often the natural outcome. By examining the data on many *sim-pua* marriages, Wolf determined that intimacy in childhood association before age three was most closely associated with divorce and lowered fertility.

A second source of data on childhood association and sexual aversion derives from ethnographic work on the traditional Israeli kibbutz, where boys and girls were raised in cohorts in children's houses, under the care of professional *metapelets*, or caregivers. From infancy through school years the children were in intimate association, being bathed, dressed, and fed together. There was no formal prohibition against marriage or sexual relations, but the children who were raised together felt more sibling-like. Melford Spiro, one of the major ethnographers of kibbutz communities, states that "in not one instance has a *sabra* [a child born on the kibbutz] married a fellow *sabra*, nor, to the best of our knowledge, has a *sabra* had sexual intercourse with a fellow *sabra*.[26]

Exceptions to the Westermarck hypothesis may be explainable. Brother-sister marriage in the royal families of Egypt, Hawaii, and the Inca, for example, had the obvious function of maintaining the integrity of the family line, and they were heavily supported by religious and political sanctions. A more challenging exception is that of Roman Egypt, during the first three centuries C.E. According to Walter Scheidel, every fourteen years during that era the residents of each household were listed in a census. Of the millions of papyrus census forms, three hundred survive. Almost 20 percent of these marriages were between full siblings, making them among the most inbred of any known human population.[27] There is supporting evidence that the partners to these marriages were true kin (the census forms list them as "brother/sister for the same father and the same mother," and, according Scheidel, "the terms *brother* and *sister* are not being used loosely or metaphorically but in the strict and ordinary senses in which we normally understand them."[28] Moreover, the marriages were institutionalized, in that they were public and celebrated.

Jack Goody regarded this case as a watershed in the study of incest taboos: "The Egyptian material must lead us to modify generally accepted ideas about the universality of the incest taboo. . . . The ball is back in the sociological court, and the game is a matter of identifying contingencies that may override widespread tendencies."[29]

There is no doubt that the Egyptian case places limits on the universality of the sibling-incest taboo and illustrates that few universals are totally universal.

Nonetheless, these unexpected data do not wholly undermine the Wester-marck hypothesis. Walter Schiedel has analyzed the same census materials and finds that brother-sister marriage in Roman Egypt was generally between siblings who were discrepant in age, on average of approximately seven years.[30] In other words, they were generally not raised together during the first three years of life, the critical period identified by Wolf for developing intense aversion. In summary, despite some exceptions, the Westermarck hypothesis has merit. Unlike the case of color, however, it is not easy to suggest a link between incest aversion and a specific neural mechanism (the retina and the neurology of color perception). Moreover, alternate explanatory models are available, including the psychoanalytic approaches to socialization.

It's Not All Relative

This brief excursion into human universals has examined some of the issues that have engaged anthropology. Culture remains the primary focus for ex-plaining custom, and human biology and psychology still have a narrow place at the table. Thanks to research on universals, however, it is no longer possible to be an uncompromising relativist. There is just too much about the human experience that is panhuman, or at least nearly so. It is true that the "voice of the species" is inflected with different accents, in that we do, of course, speak diverse languages. In the instance of the incest taboo, as we have seen, the exceptional cases are remarkable. But what lies beneath the differences and the universals is the psychic unity of humanity and the neural biology that undergirds it.

Notes

1. Lynn B. Jorde, *Genetic Variation and Human Evolution*, http://74.125.47.132/ search?q=cache:fVcJAatcEVoJ:www.ashg.org/education/pdf/geneticvariation.pdf+ human+species+genetic+homogeneous&hl=en&ct=clnk&cd=3&gl=us (accessed De-cember 12, 2008). Genetic homogeneity suggests that we are a relatively new species and that it is reasonable to consider biological explanations for common behavior.

2. Clifford Geertz, "The Impact of the Concept of Culture on the Concept of Man," in *Man in Adaptation: The Cultural Present*, ed. Yehudi Cohen, 19–32 (Chicago: Al-dine, 1974 [1965]), 21, 31.

3. Mark Hobart, "Summer's Days and Salad Days: The Coming of Age of An-thropology?" in *Comparative Anthropology*, ed. Ladislav Holy, 22–51 (Oxford: Basil Blackwell, 1987), 39.

4. Geertz, "Impact of the Concept," 24.

5. Donald E. Brown, *Human Universals* (New York: McGraw-Hill, 1991), 130 (emphasis mine). Note that Brown's definition of a universal is the most demanding one possible, in that it admits of no exceptions.

6. Brown, *Human Universals*.

7. See Donald E. Brown, "Human Universals as Compiled by Donald E. Brown," http://condor.depaul.edu/~mfiddler/hyphen/humunivers.htm (accessed December 12, 2008).

8. Geertz, "Impact of the Concept," 25.

9. Steven Pinker, *The Blank Slate* (New York: Penguin, 2000), 220–21.

10. Clifford Geertz, "Anti Anti-relativism," in *Available Light: Anthropological Reflections on Philosophical Topics* (Princeton, N.J.: Princeton University Press, 2000 [1984]), 64.

11. Pinker, *The Blank Slate*, 55 and 421ff.

12. George Peter Murdock, *Social Structure* (New York: The MacMillan Company, 1949), 2. Thus Murdock's famous definition of the nuclear family is a social group formed by marriage and "characterized by common residence, economic cooperation, and reproduction." Though based on a cross-cultural study of 250 societies, other anthropologists have preferred to see the most basic human social group as the one formed by the tie between a mother and dependent children.

13. Verne F. Ray, "Techniques and Problems in the Study of Human Color Perception," *Southwestern Journal of Anthropology* 8 (1952): 251–59.

14. Marshall Sahlins, "Colors and Culture," *Semiotica* 16 (1976): 1–22. Recent treatments of Berlin and Kay's study have introduced some complications and exceptions. A brief overview, in the context of the relativism/universalism debate, is Paul Kay's "Linguistics of Color Terms," in *International Encyclopedia of the Social and Behavioral Sciences*, ed. Neil J. Smelser and Paul P. Baltes, 2248–52 (New York: Elsevier, 2001).

15. Rodney Needham, ed., "Remarks on the Analysis of Kinship and Marriage," in *Rethinking Kinship and Marriage* (London: Tavistock, 1971), 25.

16. See especially Claude Lévi-Strauss, *The Elementary Forms of Kinship Structure* (Boston: Beacon Press, 1969 [1949]).

17. Mark T. Erickson, "Evolutionary Thought and the Current Clinical Understanding of Incest," in *Inbreeding, Incest and the Incest Taboo: The State of Knowledge at the Turn of the Century*, ed. Arthur P. Wolf and William Durham (Stanford: Stanford University Press, 2005), 162, 167–68.

18. Arthur Wolf, introduction in *Inbreeding, Incest and the Incest Taboo*, 2.

19. William H. Durham, "Assessing the Gaps in Westermarck's Theory," In *Inbreeding, Incest and the Incest Taboo*, 121–38, especially 128ff.

20. Durham, "Assessing the Gaps," 133.

21. Jonathan Haidt, "The Emotional Dog and Its Rational Tail: A Social Intuitionist Approach to Moral Judgment," *Psychological Review* 108, no. 4 (2001): 814.

22. Pinker, *The Blank Slate*, 270.

23. Cited in Wolf and Durham, *Inbreeding, Incest and the Incest Taboo*, 80.

24. Arthur P. Wolf, "Childhood Association, Sexual Attraction, and the *Incest Taboo*: A Chinese Case," *American Anthropologist* 68, no. 4 (1966): 884.

25. Wolf, "Childhood Association," 883–98.

26. Melford P. Spiro, *Children of the Kibbutz* (New York: Schocken Books, 1965), 347.

27. Walter Schiedel, "Ancient Egyptian Sibling Marriage and the Westermarck Effect," Pp. 93–104 in Wolf and Durham, *Inbreeding, Incest and the Incest Taboo*, 94.

28. Schiedel, "Ancient Egyptian Sibling Marriage," 96.

29. Jack Goody, *The Oriental, the Ancient, and the Primitive: Systems of Marriage and the Family in the Pre-industrial Societies of Eurasia* (Cambridge: Cambridge University Press, 1990), 338.

30. Schiedel, "Ancient Egyptian Sibling Marriage," 99.

10

Religion, Science, and Cognition: Explorations in Pluralistic Integration

Gary Jensen

O NE OF THE MOST-PUBLICIZED DEVELOPMENTS in biological science over the last two decades has been the "astonishing hypothesis" that phenomena popularly known as *consciousness*, the *mind*, or the *soul* have a physical existence in the brain as complexes of neurons and neural circuits, acting in ways that can be mapped and measured.[1] Moreover, because neural activity can be measured, many neuroscientists and others believe that the techniques used to discern patterned variation in the neurological correlates of consciousness may also be used to identify the location of moral and religious mental processes in the brain.[2]

Although generating considerable excitement and debate, neuroscience has yet to address a very fundamental issue: even if neuroscientists were to fully map the brain and isolate the location of consciousness, spirituality, the soul, or the mind, the cognitive "content" of those entities or processes may be as wide-ranging as Islamic fundamentalism and superstring theory. Neuroscientists acknowledge that the environment affects mental processes and that variations in the brain can affect perceptions of the external world. However, the major problems facing humankind are battles over the content of human minds. Finding the physical seat of spirituality reveals little about the diverse forms that spirituality can take, and it does not decode the variations in styles of thinking that come to characterize human beings over time, across territories, and among subgroups within society. In short, the neuroscientist's quest for consciousness has yet to lead to any testable hypotheses about the social and cultural content of consciousness.

Sociology seems an obvious academic home for theory and research about that content, how and why it varies, and how it is linked to the external social and cultural environment. However, with the exception of a few scholars who classify themselves as specialists in cognitive sociology and two quite-disparate books published over twenty years apart, cognitive sociology has yet to develop into a major specialty within sociology.[3]

Eviatar Zerubavel's excellent book outlining a cognitive sociology is considered a "field-defining primer,"[4] in which he clearly specifies the general parameters of that subdiscipline:

> Cognitive sociology reminds us that we think not only as individuals and as human beings, but also as social beings, products of particular environments that affect as well as constrain the way we cognitively interact with the world. . . . What goes on inside our heads is also affected by the particular thought communities to which we happen to belong. Such communities—churches, professions, political movements, generations, nations—are clearly larger than the individual yet considerably smaller than the entire human race.[5]

Although Zerubavel may have generated a field-defining primer, we repeat that it is a field that has no specific intellectual or substantive home in formal sociology. There is no cognitive sociology section in the American Sociological Association, and, other than a few articles on collective memory, the field has not generated a large body of research in major journals. Indeed, a search of the Internet rarely (if ever) yields sociology as one of the "interdisciplinary" fields of study comprising the cognitive sciences.

Several explanations for the relative absence of sociology from the cognitive sciences are possible: for one, the field was originally initiated by a scholar, Aaron Cicourel, who advocated a research tradition that, intentionally, did not comply with natural scientists' definitions of science. Edward Rose's review in *Contemporary Sociology* described Cicourel's approach to cognitive sociology as the study of "situated meaning," using "situated interpretive procedures."[6] Cicourel was an advocate of "ethnomethodology" as "the means whereby a cognitive sociology can be achieved."[7] The concept had been invented by Harold Garfinkel and was defined as the study of ways in which people construct and make sense of their world.[8] The focus was on "social actualization of thinking," with ideas and concepts emerging out of the study of people in interaction.[9] The approach yielded insights into rules that people constructed in interaction and negotiation with one another but did not aspire to developing insights that could be systematized and used in other forms of analysis.

Zerubavel takes a different approach to the development of cognitive sociology, in that his intention was to create "a general sociological framework for dealing with cognitive matters."[10] He intended his work to fill what he considered to be gaps in the cognitive sciences. Although he does not reject quantitative methodologies, he demonstrates his basic arguments by using examples of historical, cultural, and subcultural variation in "perceiving, attending, classifying, assigning meaning, remembering, and reckoning the time."[11]

Zerubavel has provided a primer outlining paths for a cognitive sociology as the study of "social mindscapes," and in this chapter we attempt to progress a few additional paces along that path. We begin by identifying an empirical issue amenable to quantitative analysis that can be used in the "bit-by-bit," "integrative," or "explanatory" pluralism envisioned by Kenneth Kendler.[12] According to Kendler, "Explanatory pluralism hypothesizes multiple mutually informative perspectives with which to approach natural phenomena. Typically, these perspectives differ in their levels of abstraction, use divergent scientific tools, and provide different and complementary kinds of understanding."

Such a proposal typically elicits warnings about "reductionism" and the fallacy of generalizing from findings at one level of analysis to another.[13] Indeed, such concerns are taken so seriously that there is very little, if any, research attempting to test hypotheses across scales. Sociologists are so wary of avoiding the "ecological fallacy" (making inferences about patterns of individual behavior from results of analysis at the ecological level) that they rarely carry out analysis at more than one scale.

Yet the proper response to such warnings about reductionism and ecological fallacies is to conduct research across levels on specific issues amenable to testing at different scales. Of course, it is no easy task to identify a hypothesis that can be examined in such a manner. But here we propose that one issue amenable to empirical inquiry at a variety of scales and across disciplines using different scientific tools involves variation in the importance of religion and the link between such variation and cognitive styles.

A "Bit" of Pluralistic Integration

The first step in this pluralistic integration involves a key issue in the study of religion—the debate over *secularization*, a purported decline in the institutional and personal importance of religion over time. As noted in the summary of literature following, one prominent theorist in the sociology of religion proposes that secularization involves more than a decline in the importance or authority of religion in a society. It is purported to involve a shift in "cognitive styles." Such a claim opens the door for empirical research on variations over time and space in "how people think," a promising topic for cognitive sociology.

Moreover, there are two theories focusing on the functions of religion—one psychological, with explicit ties to the cognitive sciences, and one sociological. Psychologists conducting research in the cognitive sciences propose that religion has had survival value as a means of "managing terror" and have tested this theory experimentally. A similar theory in sociology was proposed by Jack Gibbs, who suggested that variation in "perceived-control capacity" helps explain variations among societies and over time in "supernaturalism." The psychological version focuses on surges in religiosity linked to terror, while secularization and perceived-control theories tend to focus on long-term variations and variations among societies in religion or supernaturalism. Integrating the two approaches may help explain both short-term and long-term changes, as well as variations among societies at specific times.

Conceptual Convergence

Although there is a great variety of definitions, there is some agreement that religion involves beliefs or practices about "the supernatural" and some conception of a God or gods. For example, sociologist Rodney Stark defines religion as "explanations of existence based on supernatural assumptions and including statements about the nature of the supernatural and about ultimate meaning."[14] He also argues that most religions define gods as "supernatural beings having consciousness and desire" and relegates "impersonal conceptions of the supernatural" to the realm of "magic."[15]

The limitation of the concept of religion to enduring, organized, institutionalized systems of cultural beliefs and practices involving a God or gods has led some scholars to prefer the term *spirituality* to *religion*. Spirituality emphasizes personal feelings and practices, regardless of its organizational (or nonorganizational) form.[16] Yet others have included in their definitions of religion impersonal powers that Stark categorized as magic. Steve Bruce defines religion as "beliefs, actions, and institutions predicated on the existence of entities with powers of agency (that is, gods) or impersonal powers or processes possessed of moral purpose (the Hindu notion of karma, for example) that can set the conditions of, or intervene in, human affairs."[17] Neo-pagan and Wiccan groups reject the distinction made between magic and religion and the distinction made between nature and supernature. The distinction between nature and supernature breaks down when nature is depicted as divine.[18]

In *Varieties of Religious Experience*, one of the founders of psychology, William James, initiated the conception of religion as a multidimensional phenomenon,[19] and in 1950 another psychologist, Gordon Allport, proposed a distinction between "extrinsic" religion, involving behavioral practices, and "intrinsic" religion, involving beliefs and feelings. In sociology, the attempt to specify dimensions of religiosity was tied to the attempt to measure it using survey methods, with Charles Glock differentiating among beliefs, practices, experiences, knowledge, and consequences and Gerhard Lenski delineating among dimensions of the "religious factor."[20] Both used survey data to test a wide range of hypotheses about variations in religion as a multidimensional concept.

The most recent development facilitating a pluralistic integration of psychological and sociological perspectives is the cognitive science of religion. That specialty's origin has been traced to 1990 with Lawson and McCauly's *Rethinking Religion: Connecting Cognition and Culture.*[21] As summarized by Slone, the conception of religion in that field is very similar to that expressed by Glock, Stark, and other sociologists. For example, in his summary of the new specialty, Sloane defines religion as "the set of actual religious concepts people have in their heads and the behaviors they perform."[22] He also proposes that religion is a multidimensional concept that includes beliefs about supernatural agents, ritual actions, ethics, in-group/out-group distinctions, afterlife beliefs, metaphysics, and theologies. The central feature is "the belief in supernatural agents."

The conception of religion in the cognitive science of religion may be nearly identical to that used by sociologists, but there is a difference in the central questions characterizing the two disciplines: Slone identifies the central interest among cognitive scientists of religion as "explaining why most people in most places at most times have strikingly similar types of religious thoughts in their heads."[23] He identifies the central goal to be "to connect the recurrent features of religion that anthropologists and historians of religion have documented across cultures and eras to the cognitive processes involved in their production and transmission."[24] While the focus in the cognitive sciences tends to be on the recurrent features of religion, the central interest in sociology has been variations in religion over time and space and among people, along with the explanation of those variations.

Pluralistic Integration and Causality

Theories about variation in the authority or importance of religion among societies and categories of people over time tend to treat religion as the dependent variable—that is, variations in religion are presumed to depend on variations in modernization, perceived control, terror, the authority or importance of science, and other variables. Of course, the causal order could be the opposite, with the weakening of religion setting the stage for modernization.

When hypothetical independent variables cannot be manipulated experimentally, sociological research relies on the presentation of evidence (1) that there is an association of some kind, (2) that the association is not due to a shared connection with other variables (spuriousness), and (3) that supportive data of some kind justifying the assumed order of causation can be provided. Causality is not established conclusively. Rather, it is *approached* by eliminating alternative explanations and weaving bits and pieces of evidence together.

A major advantage to considering scales and hierarchies in testing theories is that evidence at one level may provide convincing evidence of causation, while evidence at other levels may provide convincing evidence that the underlying theory applies to real-world variations and patterns. For example, if an experiment can show that people or groups subjected to external threats are subsequently more likely than controls to exhibit religious or supernatural beliefs or behavior, then terror-management arguments gain in plausibility. Scott Atran has summarized experiments that support terror-management theory, showing that "the strength of belief in God's existence and in the efficacy of supernatural intervention is reliably stronger after exposure to the death prime."[25] If variations among or within nations over time were to support the same argument, then the theory would be supported under laboratory and real-world circumstances. In short, the application of methods identified with one discipline may yield patterns that might be observed at levels where causation is more difficult to establish.

Linking Historical Variation in Institutional Authority to Cognitive Styles: Macro and Micro Secularization

When it is defined as *a decline over time in religious practice or identities due to modernization,* Rodney Stark has declared secularization theory to be a fabrication by antireligious academics that ought to be allowed to "rest in peace."[26] Peter Berger appears to share that perspective and declares that "the assumption that we live in a secularized world is false" and that "the world today, with some exceptions . . . is as furiously religious as it ever was."[27]

Despite these declarations, secularization theory is not resting in peace, and two quite different theorists argue that it is either an ongoing reality or an empirical fact. In a critique of the rational-choice perspective underpinning Stark's work (involving several colleagues), Steve Bruce argues that religion is declining in significance in Western Europe and that, if Stark's argument applies at all, it would apply primarily to the United States. The polarization of advocates of the two perspectives is reflected in the title of Bruce's work *God Is Dead* and Stark's call to let it R.I.P.

Yet another line of argument is Christian Smith's contention that secularization has, in fact, proceeded in the United States but that it is not a general product of some general societal process (i.e., modernization). Rather, Smith contends that secularization was a product of an intentional revolution by opponents of religion. He states that "the historical secularization of the institutions of American public life was not a natural by-product of modernization; rather it was the outcome of a struggle between contending groups with conflicting interests seeking to control social knowledge and institutions."[28]

The central focus in this debate is whether religion has been declining, persisting, or growing in significance over one or another temporal scale. This issue is not easy to adjudicate, in that there is little consensus on the exact meaning of the concept. Stark summarizes it as the predicted "end of religion" as "humans outgrow belief in the supernatural." Numerous scholars have proposed more complex definitions, emphasizing a more structural and institutional version of secularization.[29] One of these approaches has been labeled *neosecularization* and is reflected succinctly in Tschannen's statement that "what has changed is not so much the individual's relationship to religion as the position of religion within the social structure."[30] Chaves defines it as "a decline in religious authority."[31] Stark and Finke use the concept of desacralization as identical to the macro form of secularization, where it refers to growing differentiation among social institutions over time, but challenge the secularization thesis when it is applied to individual religious commitment.[32] Peter Berger proposes a similar distinction, stating that "secularization on the societal level is not necessarily linked to secularization on the level of individual consciousness."[33] Different types of measures may yield different temporal patterns.

Despite apparent agreement on the macro version of the concept, the research relevant to it is discursive and historical. Neither data on the relative authority of different institutions nor trends over the last several decades are presented. Yet Stark insists that there should be some continuing evidence of the process, stating that, "in terms of time-series trends . . . secularization is assumed . . . to be a long, linear, downward curve" and that "it must be assumed that secularization is at least 'ongoing' to an extent that a significant downward trend in religiousness can be seen."[34] Stark's statement implies that, were secularization theory correct, some "ongoing" shifts in "religiousness" should be found. Of course, he is assuming that they will not be found. His argument implies that analyses over both long and short spans of time are potentially relevant to the identification of trends (or their absence). Absence of a trend may indicate that the process cannot be discerned in the short-run, that it has ended, or that it was never there.

The debate over secularization is also a debate about the causes of—or sources of variation in—the authority of religion. The variable most often introduced historically to explain secularization is modernization, and, as Smith argues, such causal mechanisms have been "underspecified."[35] *Modernization* has been defined in terms of alleged increases in rational planning or organizations, economic development, industrialization, and technological development.[36] Bruce presents a complex model, where all of these play a role in explaining religious change.

Of course, for an empirical researcher, progress in this debate requires specifying the meaning of concepts in ways that can be measured on a variety of scales. For example, most scholars will understand the meaning of technological modernization or development. Only recently could the proliferation of modern means of communication, interaction, and entertainment (e.g., telephones, radios, television, the Internet) be measured as properties of societies and could the correlations with measures of religious variables be used to test the hypothesis that such a form of modernization will be negatively correlated with the strength of religion.

Another dimension of modernization is the proliferation and development of educational institutions. Focusing on the United States, Smith argues that antireligious or secular elites gained control over the educational system and that secularization reflected the success of that moral enterprise.[37] However, there may be more subtle ways in which increasing educational levels of the population affect religious variables. To the degree that increasing levels of education expose populations to a variety of modern ideas, some forms of religious beliefs should decline for the population as a whole. With increasing education and literacy, the probability of exposure to diverse belief systems and ideologies increases.

Following Bruce's paradigm, modernization can be defined in terms of changes in, or the proliferation of, certain cognitive styles or social mindscapes. For example, the General Social Surveys (GSS) in the United States and the World Values Surveys (WVS) have included questions about the relative desirability of certain characteristics of children. A key contrast incorporated in both the GSS and the WVS is obedience versus independent thinking. Variations among people and nations suggested by modernization and secularization theory can be examined at a number of scales. For example, among nations, the emphasis on independence relative to obedience should be a negative correlate of the strength of religion. Among individuals in the GSS, those categorized as scientists or artists should put more emphasis on independence than on obedience than do the general professional or nonprofessional respondents. Respondents high in religious commitment and members of fundamentalist faiths should put more emphasis on obedience than independence.

Linking Terror Management and Gibbs's
Control-Capacity Theory of Supernaturalism

The underlying cause of variation in supernaturalism in Gibbs's perspective is perceived control-capacity theory, and existential anxiety plays such a role in terror-management theory. Both theories have implications for variation in religious variables as well. In fact, many (if not most) of the variables suggested by modernization-secularization theorists would have consequences for perceived control. Furthermore, Gibbs specifically predicts that variations in some forms of mortality among nations and over time help explain variations in supernaturalism, a concept that includes religion.

The terror-management theorists' use of the term *terror* shifts the focus to more specific events than to long-term changes but leads to similar predictions about existential threats and forms of religiosity or supernaturalism. For example, war and terrorist attacks should heighten "needs" for religion. High mortality rates among nations should be associated with high levels of religiosity. In short, whether the focus is on modernization, perceived control capacity, or terror management, there should be patterned, predictable variations among nations, over time, and among people.

Linking Bits and Pieces

The remainder of this chapter explores the promise of complementary findings across scales, hierarchies, disciplines, and methods. First, we consider the modernization/control hypothesis through a cross-sectional examination of variations among nations. The central question in that analysis is whether religion is less authoritative or salient in the lives of people in relatively modernized nations and whether attitudes toward science are more positive in modernized nations. Second, we must ask how confidence in religion and science within the United States has varied over time. Has confidence in religion declined relative to confidence in science and other institutions over the last several decades? Do the temporal patterns provide any support for terror-management theory? That is, are surges in external threats accompanied by surges in the salience of religion? Although no one piece of evidence can provide crucial support for the overlapping ideas summarized above, the full array of bits and pieces across scales and hierarchies may provide a coherent and convincing story.

Modernization and Religiosity among World Values Survey Nations

Although data are not available for assessing change in the authority accorded religion and modernization over time among nations, the synchronic associations using nations as units of analysis can be examined. World Values Surveys have been conducted in a growing number of societies since 1993 and have come to encompass a majority of the world's population. The surveys ask a variety of questions about religion, allowing an assessment of the relationship between measures of the control capacity or some form of modernization.[38]

Because there is considerable debate about the exact meaning of *modernization*, this analysis attempts to cover a range of possible indicators, including medical, educational, and technological modernization or development and the overall quality of life in a nation. If correlations were consistent across multiple and diverse indicators, it would be difficult to claim that modernization or control capacity has been mismeasured.

Table 10.1 summarizes the correlations across four measures of religiosity using seven different measures of modernization. All correlations are consistent with modernization/control perspectives, and all are statistically significant at the .01 level. Medical, educational, and technological development are negatively correlated with the salience of religion when measured by questions about different aspects of religiosity.

TABLE 10.1
Religious Items by Measure of Modernization

	Raised	Comfort	Confidence	Importance
Doctors per Capita	−.657	−.530	−.684	−.705
Economic Development	−.688	−.610	−.757	−.607
Quality of Life	−.612	−.735	−.880	−.780
Education	−.486	−.634	−.518	−.678
TV per 1,000	−.514	−.674	−.729	−.630
Global Technology	−.537	−.730	−.851	−.626

Notes:
Raised. "Yes" in response to "Were you brought up religiously at home?"
Comfort. "Yes" in response to "Do you find you get comfort and strength from religion?"
Confidence. "A great deal" in response to "How much confidence do you have in the church?"
Importance. "Very" in response to "How important is religion in your life?"
Definitions and sources for each measure of modernization can be found in the Global dataset in the Microcase Data Archive, Thomson Learning, Inc.
All responses are statistically significant at the .05 level. The number of nations involved ranges between 40 and 54, depending on the measure used.

Modernization and Science among World Values Survey Nations

Unfortunately, far more attention has been paid to religion in international surveys than to science. The World Values Surveys include only one item asking the following: "In the long run, do you think the scientific advances we are making will help or harm mankind?" Across nations, the average indicating that science will help is about 57 percent, ranging from highs of 80 percent in China and Turkey to lows of 26 percent in Japan and 30 percent in South Korea.

If positive attitudes toward science are captured by the percentages viewing science as helpful, then it would be reasonable to propose that at least some versions of the secularization hypothesis imply that measures of modernization should be positively correlated with a perception of science as helpful. Table 10.2 includes the correlations between percentages viewing science as helpful and the same measures of modernization as in table 10.1. The data do not support the notion that positive attitudes toward science are positive correlates of measures of modernization. In fact, while the relationships are not as strong as those found for most measures of religiosity, they are all negative and statistically significant. In short, modernization is a negative correlate of measures of religiosity as well as the measure of a positive attitude toward science. This pattern does not support the notion that measures of religiosity are weak where positive attitudes toward science are strong.

One possible explanation for this pattern is that modernization is associated with growing skepticism or cynicism, cognitive attitudes that might be reflected in attitudes across institutions. For example, the WVS asks respondents about confidence in their government. As summarized in table 10.2,

TABLE 10.2
Positive Attitudes toward Science and Government by Modernization

	Science	*Government*
Doctors per Capita	−.280	−.278
Economic Development	−.358	−.393
Quality of Life	−.512	−.347
Education	−.396	−.367
TV per 1,000	−.460	−.362
Global Technology	−.617	−.460

Notes:
Science. "Will help" in response to "In the long run, do you think the scientific advances we are making will help or harm mankind?"
Government. "Great deal" in response to "How much confidence do you have in national government?"
Definitions and sources for each measure of modernization can be found in the Global dataset in the Microcase Data Archive, Thomson Learning, Inc.
All data are statistically significant at the .05 level. The number of nations involved ranges between 40 and 54, depending on the measure used.

measures of modernization are significantly related to confidence in government, but all of the correlations are the same as those found for religion and attitudes toward science. The greater the modernization of a nation, the less confidence expressed for religion, government, or science. Thus, at a cognitive level, modernization could be defined as the proliferation of skepticism and intellectual independence.

Although the data are consistent with Smith's argument that education is the mechanism associated with low levels of authority for religion, the findings across institutions are problematic for his perspective. Among WVS nations, education levels appear to be negatively correlated with confidence in religion, but they are also negatively correlated with confidence in government or science (at least as measured here). This consistency across institutions suggests that some process other than an elite conspiracy against religion is operating, because the same patterns are found for other institutions.

Bruce suggests that changes in cognitive styles are a product of modernization, and one version of such an explanation can be considered in this chapter. The WVS include questions asking respondents to indicate the importance of a child learning "obedience" in the home and the importance of learning independence. Although the questions do not directly measure cognitive styles, they do tap relative preferences for obedience versus independence in the preferred characteristics of children in a society. Obedience should be less important and independence more important in societies with higher levels of education and modernization.

The relation between the percentages of respondents indicating *very* to each of these items and the measures of modernization are summarized in table 10.3. All of the correlations are as expected. The greater the moderniza-

TABLE 10.3
Independence and Obedience by Measure of Modernization

	Independence	Obedience
Doctors per Capita	+.369	−.601
Economic Development	+.198	−.585
Quality of Life	+.266	−.586
Education	+.221	−.468
TV per 1,000	+.208	−.520
Global Technology	+.266	−.327

Notes:
Independence. "Yes" in response to "Is it especially important for children to learn independence in the home?"
Obedience. "Yes" in response to "Is it especially important for children to learn obedience at home?"
Definitions and sources for each measure of modernization can be found in the Global dataset in the Microcase Data Archive, Thomson Learning, Inc.
All data statistically significant at the .05 level. The number of nations involved ranges between 40 and 54, depending on the measure used.

tion, the greater the emphasis on independence and the less the emphasis on obedience. It may be that an emphasis on independence in thinking is associated with a general mistrust or skepticism about a range of institutions. This pattern is quite consistent with Anthony Giddens's argument that modernity encourages a "Promethian self adaptable to changing circumstances" in which "the narrative of self-identity has to be shaped, altered, and reflexively sustained in relation to rapidly changing circumstances of social life, on a local and global scale."[39]

Confidence in Religion over Time in the United States

The research introduced by advocates of different theories in defense of or opposition to the secularization hypothesis has primarily been discursive and historical. Neither data on the relative authority of different institutions nor trends over any span of time have been presented. Stark states that there should be some continuing evidence of the secularization process were it real, stating that "in terms of time-series trends . . . secularization is assumed . . . to be a long, linear, downward curve" and that "it must be assumed that secularization is at least 'ongoing' to an extent that a significant downward trend in religiousness can be seen."[40] This argument implies that analyses over both long and short spans of time are potentially relevant to the identification of trends (or their absence).

Rather than continue debate about long-term trends where totally contrary claims are made, we concentrate instead on issues where quantitative data are available over several decades, allowing an assessment of variation, or "change," in religious phenomena. Since 1973 the GSS has included a set of items asking adults how much confidence they have in a set of institutions, including organized religion, the scientific community, medicine, education, and others. Several questions relevant to secularization can be asked: how does organized religion fare compared to the scientific community? This comparison is central to the argument about secularization in that science is typically viewed as the institution that gained authority at the expense of or relative to religion. This span of time may be too short to reveal the patterns suggested by secularization theorists, but Stark's argument that the theory implies ongoing trends suggests that it should be apparent over three decades.

The approach taken here will be distinct from prior investigations in that the focus is on variation in religion rather than a secular trend per se. Indeed, some theorists have proposed that it would be more appropriate to focus on religious change and variation in religiosity than to assume that the only important trend is a long-term downward trend. For example, in addition to possible long-term trends, measures of religiosity can be affected by sporadic events and short-term factors. One examination of GSS and other opinion poll data have reported that confidence in organized religion appears to decline in response to church scandals but to rebound with the passage of time.[41]

Variation is also caused by threats to survival. Jack Gibbs proposes that declines in perceived capacity to control external forces and threats leads to increases in "supernaturalism" but that these are primarily long-term processes.[42] In contrast, terror-management theorists imply that surges or short-term variations in perceived external threats to members of a social system may increase religiosity of various kinds. The saying that "there are no atheists in foxholes" is a version of this type of theory. When worried about unpredictable and life-threatening circumstances, religious beliefs and practices may be especially dear to people. In short, although there may be some variables that have linear, trend-like consequences, such as modernization, other variables relevant to changes or variations in types of religiosity have to be considered.

Figure 10.1 summarizes the percentages of respondents in the GSS who are very confident in "organized" religion and "the scientific community." Those data support several observations: for one, over three decades there has been a decline in the percentage of respondents expressing confidence in organized religion; in the 1970s the average was 36 percent, as compared to 25 percent in the 2000s. There is a significant, linear, downward trend, yielding a beta coefficient of $-.47$.

In contrast, there is no linear trend for confidence in the scientific community (beta $= -.04$). In the 1970s about 44 percent accorded a great deal of respect to the scientific community, and the percentage has been stable, with an average of 43 percent indicating a great deal of respect in the 2000s. If confidence in the scientific community relative to organized religion is taken as indicative of secularization, then such a pattern can be noted despite the relatively short span of years. However, the decline in confidence in religion cannot be attributed to any upward trajectory for confidence in science, since that variable has remained relatively constant over time. The scientific community has gained relative to organized religion only because confidence in the latter has declined.

The GSS asks the same question about a variety of American institutions, and the percentages are graphed in figure 10.1 as well. There are significant and prominent downward trends in the percentage indicating a great deal of confidence in education, the medical profession, the press, and television. In short, the pattern for religion is comparable to the patterns for several other institutions in the United States. The only institution exhibiting an

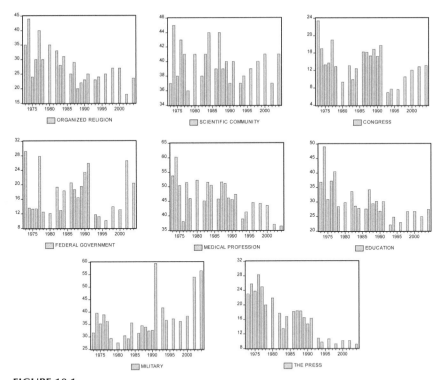

FIGURE 10.1
Percentage of Respondents Very Confident in Organized Religion and the Scientific Community. *Source:* General Social Survey

upward trend over time is the military, and that pattern is based on surges associated with the invasion of Iraq and the terrorist events of 9/11 and their aftermath.

Terror-management theory suggests that external threats are followed by increases in some form of religiosity, and that hypothesis can be tested (subject to reservations about measures) using GSS data over time. Of course, there are other variables that need to be taken into account as well. For example, the steady upward movement in educational attainment of Americans may be a source of downward movement in the authority of religion. Moreover, church scandals may have an effect as well.

Table 10.4 summarizes the results of a time-series analysis of variation in percent of GSS respondents indicating "no religion" (assigning midpoints for missing years) by percentages with at least some college experience, a dummy variable for church scandals, and two measures that might prompt the sorts of anxieties emphasized by terror-management theorists. The findings are

TABLE 10.4
Correlates of Percent of GSS Respondents Indicating No Religion, 1973–2004

Variable	Coefficient	Probability
% at least some college	+.210	.005
Church scandals[1]	+0.152	.035
Military spending[2]	–1.135	.005
External threats[3]	–0.895	.019

Notes:
Adjusted R^2 = .90
Durbin-Watson = 1.88 (With autoregressive term added)
[1] Dummy variable coding years of the televangelist scandals and Catholic Church scandals as one
[2] Military spending as a percent of GDP (from the Microcase data archive)
[3] Dummy variable coding years of the Iran hostage crisis, first Iraq War, and Afganistan-Iraq wars as one

consistent with the type of integrated model suggested in the earlier narrative. First, increases in educational attainment over time are positively related to percent indicating no religion. The more educated the public, the less religious they are. Of course, this does not mean that the majority are irreligious. Second, a measure of church scandals is associated with increases in percent nonreligious. This finding suggests that short-term as well as long-term variations are relevant to explaining changes in religious attitudes. Third, during spans of time when external threats are high and military spending as a percent of gross domestic product is high, a smaller percent are nonreligious.

Because missing years are estimated, these findings have to be considered exploratory, but they do suggest that theories addressing general trends can be combined with theories addressing short-term declines as well as short-term surges to explain variations in at least one measure of religiosity. The public is more religious when external threats are high, less religious during church scandals, and less religious as educational attainment increases. Although the causal ordering of these relationships cannot be established and inclusion of other variables might affect the model, the findings are consistent with terror-management experiments, which do establish threat as the causal variable as well as control for a variety of other variables through design. The laboratory findings are consistent with patterns found when studying a society over time.

Scientists' Attitudes about Religion and Science

As noted earlier, Rodney Stark introduces data about the attitudes of scientists toward religion to support his argument that science and religion are not in conflict. Of course, this claim is contrary to Smith's argument that scientific and educational elites have attempted to undermine religion and contrary to Ikert's statement that scientists have had to abandon the sacred for the sake of "objectivity." However, it should be noted that Stark presented data relevant to scientists only and that neither Smith nor Ikert present any systematic data at all.

The GSS does ask about occupation, and, although the number of cases in specific categories is small, the attitudes of people classified as scientists can be compared to the general population and to people in other occupations. Stark gives the impression that the level of religiosity among scientists is sufficiently high that the view that science and religion are in conflict can be rejected. Yet there may be significant differences in overall levels of religiosity and types of religious beliefs when comparing scientists to the general population. There may be variations in religiosity that fall far short of polarization and conflict.

In 1998 Hout and Greeley coded occupations to differentiate between (1) *skeptical professions*, which included natural scientists, college professors, social scientists, writers, artists, and entertainers, (2) *nonskeptical professions*, which included other professionals, and (3) all other occupations. They reported that skeptical professionals are less religious than nonskeptical professionals and all others. For example, those in skeptical professions are twice as likely to report no religious preference and attend church less often. However, as presented, the data do not allow an assessment of Stark's argument about religiosity among scientists as compared to others.[43]

By recoding the GSS occupational data, the differences among more specific categories can be examined. The responses to a variety of items dealing with religion and science among natural scientists, social scientists, college teachers, artists, and entertainers (the skeptical professions), other professionals (nonskeptical), and all other occupations are summarized in figure 10.2.

Several observations involving religion and science can be justified: (1) The percentage of respondents expressing a great deal of confidence in religion is lower for natural scientists and the other skeptical professions than for other professionals and all other occupations. (2) Natural scientists, social scientists, and college teachers have more confidence in science than do artists, other professionals, and other occupations. (3) All occupational categories express greater confidence in the scientific community than in organized religion. (4) Natural scientists, social scientists, college teachers, and artists are more than twice as likely to indicate no religious preference than are other professionals or people in all other occupations. (5) Each of the skeptical professions is less likely to attend services than other professionals and nonprofessional occupations. (6) On the other hand, there is a high level of religiosity among scientists as proposed by Stark.

A final bit of evidence relevant to the notion that education tends to be associated with skepticism is summarized in figure 10.3. In five different years

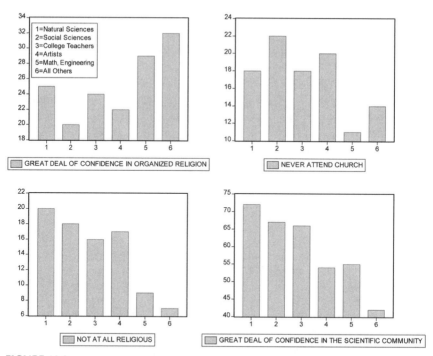

FIGURE 10.2
Answers to Religion and Science Questions from Different Professions. *Source:*
General Social Survey

the GSS asked a question about the certainty of a respondent's belief in God. About 62 percent had "no doubts," and the remainder either did not believe, did not know, or had doubts. When plotted by education and reported religious affiliation (classified as fundamentalist, moderate, or liberal), there is considerable doubt expressed by the most educated in the moderate and liberal religious categories. Indeed, most of the members of liberal denominations fall in the doubting or skeptical category. In contrast, among those in fundamentalist religious categories only a small percentage of respondents have any doubts, and there is little or no variation by education. If expressing doubts and uncertainty about God's existence is a form of skepticism, then education structures such doubts among members of those denominational affiliations classified as moderate or liberal.

Figure 10.4 summarizes the percentages of GSS respondents who choose "thinking for themselves" as opposed to "obedience" as the more-important trait for children to develop at home. As would be expected, the greater the education, the more likely respondents are to choose thinking over obedience. The liberal, moderate, and fundamentalist categories each vary as expected

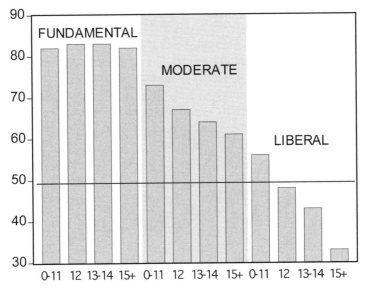

FIGURE 10.3
Religious Affiliation and Belief in God. *Source:* General Social Survey

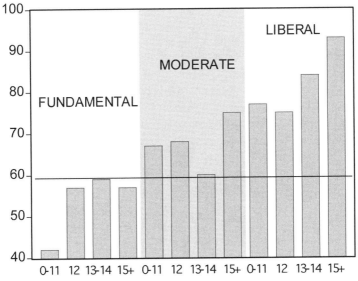

FIGURE 10.4
Important Traits for Children. *Source:* General Social Survey

as well. Among the most-educated members of liberal denominations, 93 percent choose thinking over obedience, while only 42 percent of the least-educated members of fundamentalist denominations choose that response. Moreover, further analysis shows that those who have no doubts about God's existence are less likely to choose thinking over obedience regardless of education or denominational category. It appears that independent thinking as the preferred cognitive style for children is associated with doubts about God, levels of educational attainment, and classification of religious faith as falling toward the liberal end of the continuum.

In short, the data do not support claims that scientists and intellectual elites have abandoned religion (at least in the United States), but they do suggest lower levels of religiosity, a trait shared with a variety of other professions classified as skeptical. The fact that a variety of professions share that trait suggests that it is not science per se that is associated with lower levels of religiosity but, rather, some trait shared by a variety of those professions Hout and Greeley call "skeptical." In fact, that trait may be differences in what Bruce calls "cognitive styles." However, until there is further specification and measurement of varieties of cognitive styles, that possibility is entirely speculative.

Summary and Conclusions

We have considered (1) notions of scale and hierarchy, (2) conceptions of religion, and (3) the value of bridging disciplines to overcome limitations in making claims of causality in sociology. By focusing on a basic empirical debate in sociology, the alleged decline in the authority or salience of religion in peoples' lives, analysed using available data, could be carried out across several scales. Moreover, because the focus of that debate is on variation in one or another dimension of religiosity, psychological theories linked to cognitive psychology emphasizing terror management could be integrated with sociological theories. Since tests of terror-management theory follow an experimental model, the results provide stronger evidence of causality than the type of data available for sociological analysis. On the other hand, the consistency of patterns across different types of data and analysis strengthens the external validity of the experimental results.

The notion that there has been a decline in the authority of religion appears correct only when examined independently of other social institutions and beliefs about authority. It may be more appropriate to refer to a decline in authority in general than to secularization specifically limited to religion. Indeed, a promising hypothesis for further research would be that the findings on modernization and religious authority, variations over time, and variations by educational attainment among people reflect a general shift in cognitive styles. Scientists and educational elites have not abandoned religion, but they do not accept common beliefs with the same degree of certainty as the general population. Rather than stressing obedience, their emphasis is on independence and imagination. Rather than being certain about beliefs, they are acknowledged with doubt and some skepticism. Rather than attributing a decline in religiosity to science, a promising line of argument is that increasing educational attainment tends to lead to an increase in intellectual independence and exploration of new ideas.

Although there is mounting literature and research that can be pieced together in a bit-by-bit pluralistic integration to link the study of cognition with sociological issues, the prerequisites for a step to the level of neuroscience is far more difficult. Claims that human beings are somehow neurologically hardwired to believe in God do not address variations in such beliefs. There is considerable variation among societies in the prevalence of religious beliefs of various kinds, and there is currently no research that would allow a determination that people in these different nations are neurologically different. It may be that humans are neurologically structured for abstract, symbolic, and imaginary thinking and that the location of such processes can be identified but such processes can take many different forms.

In fact, the most promising approach to integration of neuroscience would seem to be a focus on variation rather than universals. What neurological properties of the brain vary by the amount and nature of environmental input? Are different cognitive styles reflected in different neurological patterns? Do neurological patterns of children developing in a culture allowing only graphic art differ from those allowing artistic representation of humans and their gods? Is there a difference between the neurological patterns of "obedient" minds compared to "creative" minds?

The study of cognitive styles may provide a link across scales and hierarchies, and we conclude with a speculative hypothesis about cognitive styles and neuroscience. Scanning Koch's *Quest for Consciousness* for statements that might link neuroscience with hypotheses or patterns noted at other levels yields the following statement: "A brain with more explicit representations for sensory stimuli or concepts has the potential for a richer web of associations and more meaningful qualia than a brain with fewer explicit representations. Or, expressed at the level of cortical regions, the more essential nodes, the richer the meaning."[44]

It is a huge leap from this statement to proposals at other levels of analysis, but his observation prompts speculation that certain cognitive styles may be richer in webs of association than others. Are "open" minds richer in webs of association than "closed" minds? Does modernization lead to an expansion of skeptical thinking and an expansion of imagination? Does "richness of meaning" vary in such a manner that more different regions or pathways in the brain will be found to be at work for "nuanced" religious beliefs than simplistic dualisms? The pursuit of answers to such questions may allow links up the scale from neuroscience through the study of cognition to an understanding of variations among nations and changes over time.

Notes

1. F. Crick, *The Astonishing Hypothesis: The Scientific Search for the Soul* (New York: Simon & Shuster, 1994).

2. V. S. Ramachandran, S. Blakeslee, and O. Saks, *Phantoms in the Brain: Probing the Mysteries of the Human Mind* (New York: HarperCollins Publishers, 1998); V. S Ramachandran, *A Brief Tour of Consciousness* (New York: PI Press, 2004); B. E. Wexler, *Brain and Culture: Neurobiology, Ideology, and Social Change* (Cambridge, Mass.: MIT Press, 2006).

3. A. V. Cicourel, *Cognitive Sociology: Language and Meaning in Social Interaction* (New York: The Free Press, 1974); Eviatar Zerubavel, *Social Mindscapes: An Introduction to Cognitive Sociology* (Cambridge, Mass.: Harvard University Press, 1997).

4. DiMaggio, on back cover of Zerubavel, *Social Midscapes.*

5. Zerubavel, *Social Midscapes,* 9.

6. E. Rose, "Review of A. V. Cicourel: Cognitive Sociology; Language and Meaning in Social Interaction," *Contemporary Sociology* 5 (1976): 63–64.

7. Rose, "Review," 63.

8. H. Garfinkel, *Studies in Ethnomethodology* (Englewood Cliffs, N.J.: Prentice Hall, 1967).

9. Rose, "Review," 63.

10. Zerubavel, *Social Midscapes,* vii.

11. Zerubavel, *Social Midscapes,* 21.

12. Kenneth S. Kendler, "Toward a Philosophical Structure for Psychiatry," *American Journal of Psychiatry* 162 (2005): 433–40.

13. Crick, *The Astonishing Hypothesis,* 8. Crick hints at the same idea under a different rubric, arguing that "reductionism is not the rigid process of explaining one fixed set of ideas at a lower level, but a dynamic interactive process that modifies the concepts at both levels as knowledge develops."

14. Rodney Stark, *Exploring the Religious Life* (Baltimore: The Johns Hopkins University Press, 2004), 14.

15. Stark, *Exploring the Religious Life,* 10.

16. See Robert C. Fuller, *Spiritual, but Not Religious: Understanding Unchurched America* (Oxford: Oxford University Press, 2001).

17. S. Bruce, *God Is Dead: Secularization in the West* (Oxford: Blackwell Publishing, 2002), 3.

18. L. Orion, *Never Again the Burning Times* (Prospect Heights, Ill.: Waveland Press, 1995); S. Pike, *Earthly Bodies, Magical Selves: Contemporary Pagans and the Search for Community* (Berkeley: University of California Press, 2001).

19. W. James, *Varieties of Religious Experience* (Cambridge, Mass.: Harvard University Press, 1902).

20. Regarding Glock, see: C. Y. Glock, "On the Study of Religious Commitment," *Religious Education Research Supplement* (1962, July–August): 98–110; C. Y. Glock and R. Stark, *Religion and Society in Tension* (New York: Rand McNally, 1965). Regarding Lenski, see: G. Lenski, *The Religious Factor: A Sociological Study of Religion's Impact on Politics, Economics, and Family Life* (Garden City, N.Y.: Doubleday, 1963).

21. See D. J. Slone, "Cognitive Science and Religion," in *Science, Religion, and Society: An Encyclopedia of History, Culture, and Controversy*, ed. A. Eisen and G. Laderman (Armonk, N.Y.: M. E. Sharpe, 2007), 593–604.

22. Slone, "Cognitive Science and Religion," 593.

23. Slone, "Cognitive Science and Religion," 593.

24. Slone, "Cognitive Science and Religion," 595.

25. S. Atran, "Evolution and Religion," in Eisen and Laderman, *Science, Religion, and Society*, 475–83, especially 479.

26. See Bruce, *God Is Dead*.

27. Bruce, *God Is Dead*, 2.

28. C. Smith, *The Secular Revolution, Power, Interests, and Conflict in the Secularization of American Public Life* (Berkeley: University of California Press, 2003), vii.

29. J. Casanova, *Public Religions and the Modern World* (Chicago: University of Chicago Press, 1994).

30. D. Yamane, "Secularization on Trial," *Journal for the Scientific Study of Religion* 36 *(1997)*: 109–22; R. Phillips, "Can Rising Rates of Church Participation Be a Consequence of Secularization?" *Sociology of Religion* 65 (2004): 139–53.

31. M. Chaves, "Secularization as Declining Religious Authority," *Social Forces* 72 (1994): 749–74.

32. Rodney Stark and R. Finke, *Acts of Faith: Explaining the Human Side of Religion* (Berkeley: University of California Press, 2000), 200.

33. P. L. Berger, *The Desecularization of the World: Resurgent Religion and World Politics* (Grand Rapids, Mich.: William B. Eerdmans, 1999), 3.

34. R. Stark, "Secularization, R.I.P." in *Sociology of Religion* (1999), 60: 249–73.

35. Smith, *Secular Revolution*, 20–23.

36. B. R. Wilson, *Religion in Secular Society* (London: C. A. Watts, 1966); Berger, *Desecularization of the World*.

37. See Smith, *Secular Revolution*.

38. The World Values Surveys have been combined with a wide range of variables compiled by the United Nations and other agencies into a global data set archived by Microcase, available to subscribers of the Microcase Curriculum Plan. The most recent WVSs are described in Ronald Inglehart et al., World Values Study Group, World Values Survey, 1992–2002 (computer file) (Ann Arbor, Mich.: Institute for Social Research, 2002).

39. A. Giddens, *Modernity and Self-Identity. Self and Society in the Late Modern Age* (New York: Cambridge University Press. 1991), 215.

40. R. Stark and R. Finke, *Acts of Faith: Explaining the Human Side of Religion* (Berkeley: University of California Press, 2000), 59.

41. Smith, *Secular Revolution*, 360–80.

42. J. P. Gibbs, *A Theory about Control* (Boulder, Colo.: Westview Press, 1994).

43. M. Hout and A. Greeley, "What Church Officials' Reports Don't Show: Another Look at Church Attendance Data," *American Sociological Review* 63 (1998): 113–19.

44. C. Koch, *The Quest for Consciousness: A Neurobiological Approach* (Englewood, Colo.: Roberts and Company Publishers, 2004), 241.

11

The Little Divine Machine:
The Soul/Body Problem Revisited

John A. McCarthy

Radical Reality: Body/Soul, Neuroscience, Consciousness, and God

RECENT WORK DRAWING ON ADVANCES in neuroscience has taken the search for consciousness, the soul, and the so-called God gene right into the deep structures of the human central-control system: the human brain. Exemplifying this trend are Andrew Newberg, Eugene D'Aquili, and Vince Rause's *Why God Won't Go Away* and Dean Hamer's *God Gene*.[1] This hybrid organ, which bears the marks of evolutionary processes, appears to many researchers to be the ultimate frontier, for it seems as vast, as complex, and as strikingly interconnected as the universe itself. One is reminded in this regard of the Powers of 10 scaling illustrations that lead the viewer on a heuristic journey from the very, very large to the very, very small. Measuring in meters, for example, 10^{25} is equal to a billion light years (figure 11.1). On this scale, the Virgo Supercluster of galaxies appears as sprinklings of dust. Taking a view from an opposite extreme, say at 10^{-18}, the inside of a proton would also strikingly appear as a sprinkling of dust (figure 11.2). This scale is equal to 0.001 fermis, whereby a femtometer is a metric unit of length equal to one quadrillionth of a meter. Measured in time, 10^{-18} equals 1 attosecond—that is, one quintillionth (10^{-18}) of a second or one thousandth of a femtosecond. At the extremes, therefore, matter appears to be similar to the eye, the primary organ humans use to negotiate the world. Approximately midway between these extreme points on the scale of measurement is 10^0, or 1 meter (1 second in terms of time) (figure 11.3). This is the scale that we know best, that of human companionship.[2] By analogy, the human brain might be likened to the known universe that is

FIGURE 11.1
Virgo Supercluster at 10^{25}.
Source: Powers of 10 Website

FIGURE 11.2
Interior of Proton at 10^{-18}.
Source: Powers of 10 Website

FIGURE 11.3
Human Scale at 10^{0}.
Source: Powers of 10 Website

constantly in motion and has yet to reveal its ultimate secrets to the inquiring mind. The paradox of neurological research is that the object of interest is also the tool of inquiry. This paradoxical relationship will function seminally in my revisiting of the body/soul problem.

In the mid-nineteenth century, Emily Dickinson famously wrote that "the mind is as wide as the sky." While it is not one of her more felicitous verses, in context it reveals the poetess's prescience. And as a creative artist she used her mind to open up vistas that would otherwise have remained closed. This chapter is about the search for that special something that sets human-kind—situated between the extremes of the very, very large and the very, very small—apart from other animals. Whether it be hubris or simply self-confidence, humans have long believed—or felt, if you prefer—that they are more than mere animals, that they are not simply subject to the deterministic operations of the nervous system, circulatory system, and mechanically linked skeletal structure that constitute their bodies. They are convinced that their feelings have meaning beyond being instruments of physical survival.

Our understanding of the inner workings and sites of computation in the human brain is still in its infancy. Yet contemporary neuroscience has advanced light years beyond Franz Josef Gall's forensic work on the criminal mind in the early nineteenth century, in which he analyzed the brain size and composition of demented and devious individuals. Gall's method became known as *phrenology*.[3] The nascent field expanded rapidly, leading to more widespread interest as evidenced by such periodicals as the *American Phrenological Journal*, which published a version of Gall's compartmentalized version of the human brain in 1848 to explain personality traits and human behavior (figures 11.4 and 11.5).

These renditions are a far cry from the brain mappings now available in the twenty-first century, which have benefited from modern microscopy, the use of chemical stains and dyes that can selectively bind to specific molecular constituents of neurons, such as the myelin sheath that winds around axons or the ribonucleic acid, to reveal ever greater physiological detail.[4] K. Brodmann's detailed study of the physiology of the brain in 1914 represents an important step in this direction (figure 11.6). His mapping of the hierarchical structure of the cortex, which identified fifty-two discrete areas, has proved useful.[5] Yet, as simple as Gall's understanding was, it represented scientific advance over even earlier analyses. Hence, we do well to bear in mind that humans have, at every stage of knowledge growth, worked with the tools and paradigms available, adding into the mix innovative shifts of perspective enabled by new instrumentation that revealed new insights into how mind and body interact. Hence, we can trace a trajectory over time that begins with unsupported assumptions or belief, moves to a logical philosophy of mind, and continues

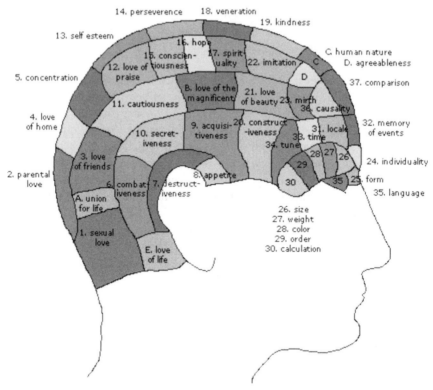

FIGURE 11.4
Hierarchy Based on Gall. Courtesy of David Pearce

with neuroscience's ability to map neural networks in an effort to locate the seat of consciousness. These are, if you will, matters of scale (each containing, of course, additional scaling). The contemporary understanding of consciousness and the older concepts of soul and mind, while apparently at odds, nonetheless begin to share common ground.

In grounding the soul in consciousness—the thesis of this chapter—I draw on a number of sources other than those already cited. For instance, in promoting his concept of *razón vital*, José Ortega y Gasset (1883–1955) advocated a conception of reason in opposition to traditional rationalism (e.g., Kant, Hegel). Howard Tuttle sums up Ortega's examination of radical reality with the following:

> Through the absorbing of circumstances, the human being constructs the world. By the use of a life form that ancient Greek philosophy designated as *pragmata*, human beings perform the actions that "must be done" for the uses and needs of human life. Such humanization of circumstances allows us to develop our

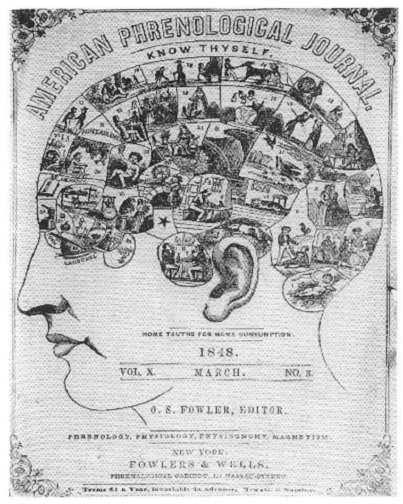

FIGURE 11.5
Journal Cover, March 1828. Courtesy of David Pearce

lives in the "system of importances" that surrounds us. Our life is a continual reabsorption of circumstances, a continual working out of our lives through them. In this respect, human beings are doomed to achieve themselves through something else, to become human through the absorbing of the nonhuman.[6]

Remarkable in the current context is Tuttle's emphasis not only on the interconnections between animate and inanimate, between the human and nonhuman, but also on the qualitative aspects of human existence that give rise to meaning. (The latter will later emerge in regard to the notion of *qualia*.)

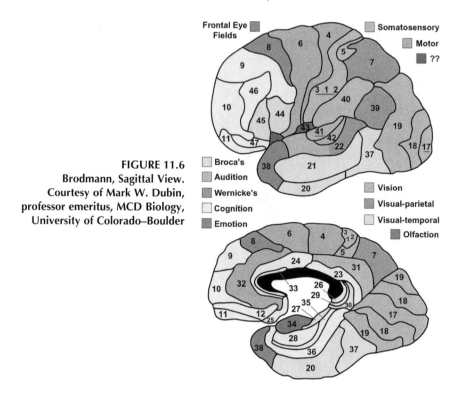

FIGURE 11.6
Brodmann, Sagittal View.
Courtesy of Mark W. Dubin,
professor emeritus, MCD Biology,
University of Colorado–Boulder

The new conception bears similarities to the vitalistic approach of Friedrich Nietzsche (1844–1900), whom Tuttle does not consider but whose famous will to power lies at the core of the human being and the world she inhabits. While for Ortega the object of study was human life as radical reality—because the human being filters and focuses all others forms of reality, much like a prism, relating those other realities to its own human one—Nietzsche's will to power applies equally to inanimate as well as animate and human nature. Nietzsche adopted a less-pronounced anthropocentric point of view.[7] In the following, I use *radical reality* in an expanded, holistic sense, one that places the body at the center of my deliberations but does not privilege the inquiring human mind as an anomaly in the physical context of things. I follow in the footsteps of those researchers and scientists who seek consilience, the unity of all knowledge.[8] But my purpose is not to engage in the kind of neurotheology that is popular in some quarters, nor is it to posit a metaphorical God gene, although I will argue that humans inherit a predisposition to be spiritual, to reach out beyond physical manifestations themselves in a quest for a larger meaning of life on earth. The latter, I suggest, is intimately interconnected with consciousness and with feelings.

Yet, in agreement with Leopold Stubenberg's position, I argue that we must distinguish between consciousness and the consciousness of consciousness (or reflective self-awareness). While introspection plays a role in the latter, it is not a factor in mere consciousness that is not self-reflective. Stubenberg clearly demarcates the two, considering their conflation in many quarters as a mere possibility unsupported by logic or science.[9] Christof Koch's recent analysis of the brain states sufficient for conscious sensory perception—what he labels *neuronal correlates of consciousness*, or *NCC*—also posits an essential difference between mere consciousness and self-knowing (*autonoetic*) consciousness that proves seminal to my premise of an intimate connection between anima and the conscious mind. He will sometimes speak of *extended consciousness* in referring to the latter, for it involves selective attention and at least the short-term storage of information.[10] Consciousness in this more-restricted meaning of deliberative, autonoetic judgment—which uses an iconic shorthand executive-summary approach to the filtering of the vast amount of visual, aural, and tactile data entering the neural circuits to pursue its evolutionary purposes—functions centrally in my revisiting of the soul/body problem.

We know all too well that what at first appears to be simple and straightforward is, on closer scrutiny, revealed to be highly complex. For instance, I grew up believing that God is Simple and that God is the Eternal Present. Meanwhile, scientific progress and knowledge expansion has taught me that nothing is simple and that all creatures big and small live in the present, too. But that similarity does not allow us to conclude that they are God or, as Baruch de Spinoza put it more than three hundred years ago, *natura sive deus*. The present-time warp of these creatures large and small across the entire scale of things, from the microscopic to the meson and cosmologic plane, is contextualized by past experiences. And the mode of living in the present is conditioned consciously or unconsciously by a desire to live for another day. God—as the devout believe—exists in the present, too. But all things past, current, and future are eternally coexistent in the mind of God. If this were not so, then God would not be omnipotent and omnipresent, qualities that she must have to qualify as absolute and genuinely divine. All earthly creatures are subject to memory, memory loss, and the vagaries of time and place. God is not.

In revisiting the problem of the body/soul/mind problem, we focus sharply on the following: God's awareness of past, present, and future events is an eternal present, essentially different from humankind's experience of the temporal modes, which they see as a sequential continuum including memory and projection. Strikingly, the temporal lobe, the primary site involved in long-term memory and emotion, is associated with religious experience. And

God would of course have the longest-term memory as the eternal present. Yet via the temporal lobe humans do have a window into the eternal, even if the viewing glass is smudged and distorted. Questions of hierarchy and scale are central to how I intend to relate the eternal God-present to traditional notions of the soul and mind, and to more contemporary explanations of mind and consciousness. Equally seminal to my argument is the modern scientific understanding of the inherent relationship between energy and matter. I will speak of body/soul in terms of matter and energy. Much of what I will argue is speculation. And I will not apologize for it, as even hardcore neuroscientists and mathematicians such as Christof Koch, Eric Kendall, and Roger Penrose do not shy away from speculating about what happens at the neuronal scale (where interactions among nerve cells occur on the millisecond time scale in equally miniscule, circumscribed space dimensions) when their validation methods for the inner workings of mind and consciousness exceed their quantifying grasp.[11]

Finally, a third pivotal feature of my argument is the essentiality of networking, connectionism, feedback loops, and the constant testing of relationships among the parts of any given system and among various interacting systems. We do well to think in terms of a grandiose ecological network of systems that have produced the biological systems of Mother Earth and, hence, of the human species with its hybridity of animal needs and spiritual projections. *Spiritual* is intended in the German sense of *Geist*, which includes the Roman Catholic *Heiliger Geist*, the Goethean *Erdgeist*, the Hegelian *Weltgeist*, and the Ranckean *Zeitgeist*. In each case, interconnectionism is the operative principle. This we recognize in the Roman Catholic catechism's explanation of the origin of the Holy Spirit: the Holy Ghost emanates from the exchange of energy between God the Father and God the Son to form the Holy Trinity. The Holy Spirit is thus a kind of bonding agent. I utilize the phenomenon of energy exchange, whether we encounter it in its materialized or pure form.

Much progress in the scientifically objective analysis of the world we inhabit and the body we need to negotiate it has been made in recent memory. The study of the brain and its neuronal structures has emerged of late as the most-promising area of investigation for fathoming the nature of human consciousness (or of self-awareness). For instance, we now know that information generally emanates from the back of the brain to the front and that the brain expends most of its energy packaging sensory input from all available sites to create a manageable view of the environment. Visual and somatosensory data give us a sense of where are bodies are in space. Memory functions in the temporal lobe allow us to recognize the visual perceptions. Finally, all this processed input is sent to the frontal lobe, where judgments are made about what action should be taken.[12]

Symptomatic of this mapping on the nanoscale is Christof Koch's *Quest for Consciousness: A Neurological Approach* and Roger Penrose's *Shadows of the Mind*. Koch emphasizes the hierarchical structures and grouping of neurons in the cortical areas at the rear of the brain that evince at least a dozen levels in his quest for understanding how consciousness arises.[13] He cites the mass of two hundred million axons in the corpus callosum, which relays sensory or symbolic information from one side of the brain to the other, as just one example of the mediatory processes in the rise of consciousness, which we are just barely beginning to understand.[14] On the other hand, the mathematician Penrose speaks of cytoskeletons that consist of protein-like molecules arranged in various types of structures, such as actin, microtubules, and intermediate filaments. Of particular interest to him are the microtubules, protein polymers in the shape of a hollow cylindrical tube, some twenty-five nanometers in diameter on the outside and fourteen nanometers on the inside. Each consists of dimmer subunits known as *tubulin*, which in turn consists of two distinct parts. Each of these, again, in turn, is composed of about 450 amino acids. There are normally thirteen columns of tubulin dimmers forming the external walls of the hollow tube of each microtubule. In absolute units, the mass of each microtubule is about 10^{-14}. Microtubules, existing at the margin of the quantum and classical scales of interaction, are sometimes organized into larger tube-like fibers. Penrose sees the roles played by these elastic structures as promising for explaining the physical foundations of consciousness.[15] Andrew Newberg, Eugene D'Aquili, and Vince Rause have sought to trace the so-called God gene in their book on *Why God Won't Go Away*, placing heavy emphasis on the temporal lobe.[16] As diverse as are the approaches by Koch and Penrose, on the one hand, and that of Newberg and colleagues, on the other, all approaches share an essential assumption in common: the deep structure of the brain and its neuronal connections are the key to explaining the origins of consciousness (for Koch and Penrose) and for the human penchant to believe in God and the human soul (for Newberg et al.).

This point is manifest in their respective hierarchic organizational modeling of the brain with its primitive inner limbic system, the surrounding layers and sections of the neocortex, and the intricate web of constantly morphing neuronal pathways throughout the entire tripartite structure. Koch discusses the thirty or so processing areas of the brain specifically involved in vision (his thematic focus). These functions include various parts of the hindbrain, midbrain, and forebrain. They, in turn, are organized into further reductionistic components (e.g., cerebellum, occipital lobe, corpus callosum, thalamus, hypothalamus, parietal lobe, frontal lobe, temporal lobe, pons, medulla). Because of the porous boundaries and many feedback loops among these processing centers, it is not possible to speak of a strict or unique hierarchy of the components. Koch also notes that similar hierarchical organizations have been identified for somatosensory and auditory regions of the brain.[17] Although Koch does label the prefrontal cortex "the seat of the executive,"[18] the hierarchical organization of the front of the cortex remains unclear. Hence, what emerges as central for Koch is not any privileging of the hierarchical centers in the brain that process raw input entering through the eye (or through the spinal cord) itself, but, rather, the neuronal pathways and the myriad feedback loops themselves that link in ascending, descending, and lateral ways and exhibit varying time durations and impulse strength. They are critical to establishing a notion of hierarchic structure. "There is no single Olympian region," Koch remarks, "that looks down on the entire visual system. . . . No region is a one-way nexus. If it were, it could not be a causal agent; it could not mediate any useful consciousness."[19] To be sure, the brain has devised an efficient way of dealing with the overload of stimuli by taking "shortcuts" to established "fixed" patterns or chains of previously experienced neuronal connections. Despite these established pathways, however, other connections and feedback loops occur regularly. These frequently and less-frequently recurring connections he relates to the NCC.[20]

Newberg's discussion of brain architecture is incomparably less anatomical than Koch's analysis (or Penrose's). Yet there is considerable agreement between the two models as to the essential functions of the brain's various components. Where they differ is in Newberg and company's readiness to speculate about the value of the individual processing centers and their ease in privileging certain components. For Newberg, the limbic system—what he calls the *emotional brain*—mediates between humankind's lower animal being and its higher mental/spiritual dimension. Its privileged position is directly related to its involvement in religious and spiritual experiences (meditation, out-of-body experience), even though its primary function is to generate and moderate primal emotions.

The three major components of the diencephalons—the hypothalamus, amygdala, and the hippocampus—are respectively labeled *master controller*, *watchdog*, and *diplomat*.[21] As these nomenclatures suggest, the hypothalamus is positioned between the animal and the spirit on its perch atop the spinal cord, reaching up into the interior of the brain. While it can help generate basic emotions, such as terror and bliss, it also receives input from the amygdalae, located in the middle part of the temporal lobe, and "mediates all higher-order emotional functions" (e.g., love, distrust). Via its neuronal connections with other brain regions, it monitors other parts of the brain for stimuli warranting special attention either because they pose a threat or because they offer an opportunity for action. In other words, Newberg avers, the amygdala assigns emotional value to stimuli and passes the information on to the hypothalamus.[22] Located behind the amygdala in the temporal lobe, the hippocampus frequently acts in concert with it. It also seems to be linked to the thalamus,[23] whose primary function is to maintain emotional equilibrium by blocking sensory input to various neocortical regions and is, hence, instrumental in regulating a person's state of mind.[24]

Having described the chief functions of the oldest parts of the brain and establishing their medial position between the lower and higher functions of the human being, Newberg turns his attention to those highly complex processing centers distributed throughout the mammalian brain. He calls them *cognitive operators* and readily admits that they are not easy to describe, for they are not *eo ipso* structures but, rather, "collective functions of various brain structures."[25] He then delineates a "holistic operator" (located in the right hemisphere), a "reductionist operator" (left hemisphere), a "quantitative operator," the "causal operator," the "binary operator," the "existential operator," and the "emotional-value operator."[26] Obviously, the qualifiers designate particular cognitive, judgmental, and qualitative functions of what Koch labels NCC. While each operator is assigned a specific analytic function in the processing of data, the emotional-value operator assumes an integrative function, assigning an emotional valence to elements of perception and cognition, which essentially raises us above the level of mere automata. Based on his insights into brain architecture, Newberg feels compelled to believe that God—if she does exist—could manifest herself "in the tangled neural pathways and physiological structures of the brain."[27]

Strikingly, Koch and Penrose would locate consciousness in these very same complex neuronal pathways with their myriad feedback loops into various processing centers. Of course, one can readily object that Newberg's analysis draws heavily on speculation. At the same time, however, Koch devotes a couple of his chapters to speculative extrapolations from physiological characteristics to functions of consciousness (e.g., chapters 13 and 18). This he does despite easily dismissing as speculative Penrose's claim that the foundations of consciousness are based on quantum-mechanical effects; Koch finds no evidence of such operations.[28]

A further bridge between the willing believer (Newberg) and the ready skeptic (Koch) is Koch's fascination with qualia, those elemental feelings and sensations that make up conscious experience and constitute meaning. "Qualia," he avers, "are too structured to be an irrelevant by-product of the brain."[29] In a series of ten working assumptions about qualia too involved to recapitulate here, Koch concludes that qualia are "a property of parallel feedback networks" that, based in "the cortex, thalamus, basal ganglia, and other closely allied networks," crystallize as distinct feelings and nodes of significance.[30] Qualia are the result of enormous feedback loops that involve not only the essential nodes in a particular sensory perception but also store information and related responses in neighboring cells. Although the brain activity involved in their emergence lasts for only a limited amount of time, the activity does repeat itself. Koch does not align qualia with the NCC but rather sees them as a product of their *penumbra*—that is, "the neural substrate of past associations, the expected consequences of the NCC, the cognitive background, and future plans."[31] These "shadows of the mind" figure centrally in the ultimate thesis about the foundations of consciousness.

Qualia symbolize, therefore, a narrowing of focus based on a vast amount of tacit information in nigh iconic fashion. They are "potent symbolic representations" (what a literary scholar is tempted to call *symbols, metaphors,* and *tropes*) that are instrumental in establishing the meaning of a text. Why qualia feel the way they do remains an enigma, Koch avers,[32] but he agrees with Newberg that they contribute to our sense of control in a world not originally of our making. This very feature of constituting meaning is also central to Leopold Stubenberg's concept of consciousness. He devoted an entire monograph, *Consciousness and Qualia*, to defining qualia and the role they seem to play in self-awareness. Koch does not seem to know Stubenberg's work either, although it is entirely germane to his own theory of qualia.[33]

For my present purpose the previous means that consciousness seems to arise with the ever-greater accumulation of sensory data, which the brain processes in a highly complex fashion that is not yet completely understood. Yet "the brain, with more explicit representations for sensory stimuli or concepts, has the potential for a richer web of associations and more meaningful qualia than a brain with fewer representations."[34] In other words, the greater the number of essential nodes in the cortical regions (with their numerous feedback loops to other regions of the brain), the richer the meaning will be. Put differently, the higher the energy levels in bundled rather than dispersed manner, the richer the potential for giving meaning to stimuli. Manifestly, enhanced complexity of discernment and meaning are intimately linked. The implicit assumption here is not only the role of unarticulated data in the making of meaning but that more richly textured meaning is a good thing. It is precisely this connection and this assumption ordained as legitimate in the doing of science that allows me now to turn my attention to two early philosophers of the mind and of nature, who, in light of the foregoing, appear to have been saliently prescient: Gottfried Wilhelm Leibniz and Baruch de Spinoza. I disagree, therefore, with William Barrett's thesis about the death of the soul resultant of scientific progress.[35]

Divine Machine and Supreme Substance: Consciousness and God

Gottfried Wilhelm Leibniz (1646–1716)—courtier, mathematician, legal historian, diplomat, librarian, physicist, and philosopher—is perhaps best known as coinventor of calculus and as author of *Essais de héodizée sur la bonté de Dieu, la liberté de l'homme et l'origine de mal*, published in 1710, and the *Monadology*, published in 1714, two hundred years before Brodmann's mapping of the cortex. Of interest in our present context is Leibniz's endeavor to reconcile a theologicoteleological conception of the world with the new physicomechanical view. In the process he developed a theory of the unity of body and soul in the universe that broke with Cartesian dualism. Written for Nicolas Remond, an admirer of Leibniz's *Theodicy*, the title *Monadology* originated with Heinrich Köhler (1685–1737), a natural-law historian in Jena, who provided it in his translation of 1720 Leibniz's work, *Des Herrn Gottfr. Wilh. von Leibniz Lehrsätze über die Monadologie*. Leibniz penned a similar essay in 1714 for Prince Eugene of Savoy called *Principles of Nature and Grace, Founded on Reason*, where many of the same ideas recur. In fact, the essential features of Leibniz's view of the relationship between body and mind in a unified universe are discernible even from his *Discourse on Metaphysics* (1686), penned in reaction to Descartes' dualism of body and mind.

The *Monadology*—which, in ninety axioms, provides a kind of executive summary of the *Theodizee*—presents the essence of Leibniz's newly branded philosophical construct: in *New System* (1695) Leibniz had earlier noted that his outlook emerged after he had freed himself of the yoke of Aristotle and his belief in atoms and the void.[36] Major ideas presented in the *Monadology* include (1) the unity of the universe, (2) the concept of organism or growth, (3) the principles of activity and perfectibility, (4) the belief in evolutionary progress (which he called *optimism*), (5) the preestablished harmony of body and soul, and (6) the increasing complexity of cognitive discernment that functions as a link between the human and the divine. In the following I concentrate on Leibniz's notions of soul and mind and their interactions as expressed primarily in the *Monadology* (1714), *Principles of Nature and Grace, Founded on Reason* (1714), and *Discourse on Metaphysics* (1686). None of what Leibniz has to say, however, is based on any detailed examination of the brain itself, although by analogy his views of complex interconnectedness and the rise of consciousness are germane to contemporary scientific mappings.

The essential feature of Leibniz's philosophy is his concept of the *monad*, which combines the principles of simplicity, unity, and multiplicity in the smallest indivisible package. Drawn from the Greek word *monas*, signifying oneness and unity, the monad designates a simple substance that is capable of self-determining action and that exists in conjunction with physical attributes.[37] A simple substance is one that has no parts, unlike a machine, which consists of interconnected pieces. In *Principles of Nature and Grace*, Leibniz explains further that simple substances or unities are "lives, souls, minds." Elsewhere he labels monads "atoms of substance" and "metaphysical points" that appear to us to be mathematical points in the world of phenomena.[38] These simple substances exist everywhere in conjunction with matter; without them there could be no compounds. (Only God is without material body.) Consequently, "the whole of nature is full of life."[39] Later in the *Monadology* he uses metaphors of "a garden full of plants" and "a pond full of fish" to symbolize his conception of the living universe where there is no void, no waste, no sterility, and no genuine death except in the appearance of things.[40] What we call *death*, he proffers, is "an envelopment and diminution" of force and activity in bodies, while *generation* is a "development and growth."[41] All bodies are in a "continual state of flux, like rivers."[42] All created things—even the monads themselves—are subject to change, and, indeed, this change is continual.[43] Hence, we have the foundation of his signature feature, of vitalism, change, and interconnectionism.[44] As a mathematician and biologist, Leibniz sought to explain the phenomenon of organic life without, however, rejecting the great principle of mathematical physics that Descartes had established. "The harmony between the two ways of thinking," Cassirer avers, depends on the realization "that all the phenomena of nature, without exception, are capable of a strictly mathematical and mechanical explanation. However, the principles of mechanism themselves are not to be looked for in mere extension, shape, and motion but as issuing from another source."[45] That is where the metaphysical order of substances reenters the debate. That is what Leibniz sought to accomplish with his *Monadology*.

Because of the principle of change, Leibniz reasons, there must be "a plurality within the unity or the simple."[46] Obviously, he could not know the complexities of molecules or of biological and neuronal cells. Microscopy was in its infancy. For that reason, Leibniz argued on the basis of analogy from observations. Qualia, as they are now known, figure prominently in his ruminations. Thus he contrasts his notion of "incorporeal automata"[47] with mechanical machines, such as a millwork or a timepiece. For example, one can isolate the teeth in gears in a clock and can identify mechanical pulleys and springs and see how they impinge on one another. But one cannot identify how self-motivated change occurs using such a mechanical model. A compound or machine fails to explain metamorphosis for Leibniz. Instead, he posits that a "perception" must be sought in the monad. It is in this event of change alone, he states, "that all the internal actions of simple substances must consist."[48] In his *New System of the Nature and Communication of Substances, as Well as the Union Existing between the Soul and the Body* (1695), he had previously underscored the nature of monads as "animated points" or "atoms of substance," different from the Aristotelian conception of atoms existing in a void:

> It is only *atoms of substance*—that is to say, unities which are real and absolutely without parts—that can be the sources of actions and the absolute first principles of the composition of things and, as it were, the ultimate elements into which substantial things can be analyzed. They might be called *metaphysical points*; there is about them *something vital* and a kind of *perception*, and *mathematical points* are their *points of view* for expressing the universe.[49]

This "something vital" about the theory of monads resonates with Ortega's *raizon vital* and the notion of radical reality—that is, a reality constantly filtered by the sensory and cognitive operations of the human organism. The meaning derived from the sensory input is phrased in terms of how it impacts the human being herself. The "atoms of substance" constitute radical reality for Leibniz around 1700. What he says about the universe (*plenum*) would also apply to various brain processes; the brain itself is analogous to a monad that is multifaceted yet operates in an integrative fashion as a single biological organ.

The metamorphoses in the living substances result from an internal principle of action called *appetition*. Leibniz emphasizes the need to distinguish between these passing states (which he calls *perceptions*) and more enduring states of attuned consciousness (which he calls *apperception*).[50] The former are events in the soul/body or indivisible unit of "living substance," whereas the latter can only occur in higher animals (humans). Leibniz says, "but it is the knowledge of necessary and eternal truths that distinguishes us from mere animals and gives us *reason* and the sciences, raising us to knowledge of ourselves and God. It is this in us that we call the rational soul or *mind*."[51] Appetition—or the action principle—is directly associated with the original principle of action—God—and functions like a compass that points toward the magnetic pole. And, like a compass needle confused by conflicting magnetic points of attraction, the action can be temporarily misdirected until the interference has been overcome. The rational soul learns how to read the needle and determine True North, so to speak.

The difference between perception and apperception correlates to the contemporary distinction between automatic events in our bodies that do not rise to the level of consciousness (such as the blood coursing through our veins or the firing of neurons), on the one hand, and the consciousness of consciousness on the other (such as the sense of self struggling with an ethical dilemma). Leibniz's attempt to replace a mechanical (later LaPlacean) conception of the workings of nature with an animistic one might be seen as an early expression of complexity theory that emphasizes relations and connections to explain processing centers and bundled waves (or points) of energy. The latter Leibniz labeled *animated points.*[52] In the *Monadology*—where he transitions his argument from the operations of the individual monads, which comprise the material world, to their place within the total operation of the universe—Leibniz states that the whole of matter is interconnected, that "as in a plenum every movement has some effect on bodies in proportion to their distance" and is affected not only by direct physical interaction and first contact but also indirectly by much more distant acts and movement.[53] The monads communicate throughout the entire system. This is possible because each monad is a mirror of the entire universe. And this mirroring is possible because of the hypothesis that God is the great architect and law giver of the created world. God is the *Urmonade*, with which all created monads are related, with whom they communicate. Therefore, they share in the original creative and unifying force that is God.

God is thus the "Supreme Substance."[54] In God is power, knowledge, will (*puissance, conaisance, volence*).[55] Schneiders points out that Leibniz's conception of God as absolute being and essence—as the "region of eternal verities" and the cause of created matter—represents the epitome of the fullest possible development of latent potential in man and in the universe itself.[56] If the deity is the "region of eternal verities" (comparable to the Platonic realm of noumena), then human reason (*Verstand* and *Vernunft*) is the "region of possibilities" within the world of appearances (phenomena). Reason, then, is the instrument used by humankind to negotiate the physical environment in order to enhance its innate capacity to comprehend the whole. God is the guarantor that the plurality of perspectives and manifestations are not ultimately chaotic but are constitutive of an underlying order. She is the guarantor that human cognition is even possible. In principle, God is the most complete, most distinct, and most encompassing thought realizable. In this regard, God is "*le vray point de veue.*"[57] Because of this connection between God as creator and created world as apparently autonomous system, capable of synthesis and evolutionary development, Leibniz speaks of our environment as the best of all possible worlds, a phrase that has often been misunderstood after Voltaire so stridently mocked it in *Candide, ou l'Opstimisme* (1759).

Harmony, then, "*est similitude in dissimilibus.*"[58] What appears to us to be dissonant in the world is dissonant only because we do not yet understand how the aberration is integrated into the total system. Researchers such as Koch and Penrose readily admit that our understanding today of brain functions, the genetic code, and the expanding universe is similarly limited. Because of our knowledge is incomplete and our perceptions imperfect, Leibniz can claim that there is no chaos in the world, that we simply do not have the full picture (a picture reserved for the deity alone). Today scientists and philosophers also tend to reject the notion that we live in a chaotic world, even though we are seemingly surrounded by chaotic events.[59] For Leibniz as well as for us, harmony of body and soul can be seen as the reconciliation of two apparently negative moments in a higher organizing principle that might explain away apparent contradictions of evidence. However, this reconciliation does not connote a dominance of one moment over another but rather an agonistic coordination of seemingly opposing forces or perspectives.[60] Because there is so much that we do not know, how can we presume to assert that reconciliation of apparent opposites is not possible? Perhaps humankind has a penchant to seek reconciliation as an automaton response to the environment that produced us, or perhaps our desire for harmony and unity is nature's way of communicating with the individual organism. Perhaps the entire human organism "is geared to take advantage of noncomputable action in physical laws," as Roger Penrose surmises about the function of cytoskeletons, those structures that appear not only to give the neuronal cell shape but also to be the control mechanism for the cell itself.[61] By analogy, one could assert that the cytoskeleton is the combined unit of skeleton, circulatory system, muscles, and nervous system, all rolled into one processing unit.[62] Increasing the scale, we can link the microtubules in the neuronal network with the human body with its sensory inputs and mobility as similar kinds of computational, "animated points." Monads are everywhere working in concert toward some dimly conceived end. Teleological? Yes, but not mechanical or linear.

Within Leibniz's system there is a hierarchy of scales—and here I am concerned only with sentient beings—that lead from plants to insects to brute animals to more highly developed animals. All beings capable of metamorphosis have an entelechy—or a natural automaton—that controls change internally and is not simply dependent on external forces. The varieties of form and perception create a multiplicity of possible takes on the whole but within the unity of the whole. Every body is sensitive to what is happening in the universe. Leibniz writes, "Thus, although each created monad represents the whole universe, it represents more distinctly the body, which is particularly affected by it and whose entelechy it is; and as this body expresses the whole universe by the connection of all matter in the plenum, the soul represents the whole universe also in representing the body, which belongs to it in a particular way."[63]

The preestablished harmony between a particular soul and its particular body ensures its individualism as an organic "living thing."[64] This living thing he calls an animal. All animals have a soul, but not all souls are equally developed. All animals have perceptions, but Leibniz thinks that possibly only humans have apperceptions (he is unsure about brute animals). Hence, only humans have the ability to rise from merely sensitive souls to the rank of reason and ultimately to "the prerogative of minds."[65] While the soul acts according to the laws of final causes by virtue of "appetitions, ends, and means," bodies act according to the principle of efficient causes by motions. Yet these worlds of final and efficient causes cooperate according to a preestablished harmony, like the prongs of a tuning fork resonating in unison.[66]

Before one can have an apperception, one must perceive. In fact, we experience an infinite number of perceptions all the time, although they do not rise to the level of conscious reflection. The reasons for the lack of apperception, Leibniz surmises, might be that the impressions are either too small, too numerous, or too undistinguished to stand out. But in the mass, in combination with other sensory data, they nevertheless "do not fail to have their effect and to make themselves felt, at least confusedly."[67] In this response to John Locke's *On Human Understanding*, Leibniz draws a comparison to the sound of a waterfall that we do not consciously register once we have grown habituated to the sound (likes noises in the urban landscape). The water still flows, and the sounds still emanate from it, but we are not attentive to them. Clearly, he concludes, attention is critical to the transformation of a perception into an apperception. We might add that a bundling of energy is necessary for a standing wave to rise from the undifferentiated surface of the water. Here solutions immediately come to mind that appear in different media and on the micro and macro levels, not just in water. They are found in molecular biology as ripples of energy, in superconductors as fluxons tunneling their way quantum mechanically through the insulating barrier between the sections of superconducting material, and as the electric-charge polarons that move through solid-state crystalline structures.[68]

Humans have a soul capable of consciously reflecting on sensory input, which in turn leads to knowledge of "necessary and eternal truths" (God), which can be likened to unchanging laws of interaction not summed up in the individual moments of metamorphosis themselves. This capacity for self-reflection distinguishes the human soul as *rational*, christening it *mind*.[69] Moreover, self-reflection leads to our sense of self. "In thinking of ourselves," Leibniz remarks, "we think of being, of substance, of the simple and the compound, of the immaterial, and of God himself, conceiving that what is limited to us in him is limitless. And these acts of reflection provide the chief objects of our reasoning."[70] The inherent relationship that Leibniz posits between God as *Urmonade* and human as rational monads guarantees that humans share in the energy (*puissance*), knowledge (*conaisance*), and will (*volence*) that is God.

Obviously, the imagination and the ability to project completion on the half-formed image both play major roles in this system, as they do, of course, in contemporary science. In fact, the imagination functions for Leibniz as a guarantor that humans will explore possibilities beyond the given. As the sense of self expands, it will draw closer to the deity. And all conscious selves taken together represent a rapprochement of the multiplicity of perspectives with the godhead. Whereas the soul (monad) embedded in its body is in general a living image of the created universe, the mind is an image of divinity itself. Via its ability to know the system of laws governing the universe and of the human anatomy, and because of its facility in imitating something of those laws by architectonic patterns, each mind is essentially "a little divinity in its own department."[71] At the conclusion of the *Monadology*, Leibniz envisions the City of God as the ultimate assemblage of all minds, such that the City of God would actually be a city of minds ruled by God the monarch once the perfectibility of our mutable selves and the universe has been fully realized.[72] Hence, consciousness of consciousness "renders minds capable of entering into a kind of society with God and makes his relation to them not only that of an inventor to his machine (which is God's relation to the rest of created things) but also that of a prince to his subjects, and even of a father to his children."[73]

Minds capable of entering into communion with the divinity are the most active "animated points" or "atoms of substance." In explaining the role of evil in the world, Leibniz has recourse to the principles of activity and passivity in the monads. Evil denotes for him, above all, the divergent force of stasis and the lack of convergent dynamic interplay, as when, for example, a monad in a more passive phase is overshadowed by another, more active, one. God as perfect activity sees only the active modes.[74] For this reason Leibniz contends that there is no need for a principle of evil; it is all a matter of relationships.[75] Because of their centrality in explaining the unity of body and mind, of the material and the divine, Leibniz was compelled to rehabilitate the Platonic concept of substantial forms, which had fallen into disregard. In his *New System* he had therefore explained the driving force behind his need to reintroduce the concept, but now, in the *Monadology*, he spoke of it unambiguously as agency:

> I found, then, that their nature [substantial forms] consists of *force* and that from this there follows something analogous to feeling and to appetite and that therefore it was necessary to form a conception of them resembling our ordinary notion of *souls*. But just as *the soul must not be used to explain the detail of the economy of the animal's body*, so I judged in the same way that these forms ought not to be used to *explain the particular problems of nature*, although they are necessary to establish true general principles. Aristotle calls them *first entelechies*; I call them, more intelligibly perhaps, *primitive forces*, which contain not only the *act*, or the fulfillment of possibility, but also an original *activity*.[76]

Like so many energy pulses, all monads are related via the activity principle to the *Urmonade* or original *Strahlkraft*.[77] Because of the active role that the mind plays in Leibniz's philosophy of nature, it does not seem right to designate him as a mere rationalist.[78] For him the mind was not a blank slate on which experiences are imprinted but rather a mirror and active inscribing of experience in the manner of the creator himself. The conscious soul is a "little divine machine," quite like man's projection of the nature of God herself. The more active the mind, the more god-like it is. And because Leibniz's mapping of reality defines God as the principle of action, there is the implication that the search for completion or the expansion of knowledge will not result in stasis. While individual energy knots (or animated points) might dissolve into the surrounding context, the whole will continue realizing the multiplicity of possibilities designed into the operations of the universe on all scales, from the very, very small to the very, very large.

The biblical Garden of Eden contained the seed of contemporary consciousness in the forbidden fruit of the Tree of Knowledge of Good and Evil. Only by means of eating the apple, however, could that seed bear full fruition, as Baruch de Spinoza argued. It enabled a new mode of existence, one that lends fullness to human nature according to the Intelligent Design as understood by Leibniz. Expelled from paradise, Adam and Eve's destiny is to struggle through the multivalent possibilities for development outside Eden in an effort presumed to lead back to the well-tempered life in the wholeness of God. That way, however, is blocked; the path leads forward to ever greater complexity. The chaotic bifurcations of the roots and branches of both the Tree of Life and the Tree of Knowledge resonate with one another in telling fashion from the outset.

The dynamic interplay of bifurcation and period doubling of life in nature make it almost impossible to fulfill the divine exhortation to return to undividedness. Hence, eating the forbidden fruit marks the beginning of self-awareness, diversity, and, ultimately, potentiality. From simplicity, complexity arises. The process is godly. To speak with Holmes Rolston, "God created earth as the home (the ecosystem) that could produce all those myriad kinds. . . . There is nothing ungodly about a world in which every living thing defends its intrinsic value, those brought forth from its own perspective, at the same time that it shares, or distributes, these to offspring, to others, in the ongoing evolutionary narrative."[79]

In human terms, the evolutionary narrative began with Adam and Eve's mythical fall to the ground. The *biophilia* hypothesis proposed by Edward O. Wilson and Stephen Kellert seeks to explain the recurrent need of modern urban humans to return to nature, be it in the city park, the suburban back yard, or the open countryside.[80] The phenomenon of biophilia is a sign of the interconnectedness of all things. Arthur Lovejoy, following in the footsteps of Leibniz, acknowledged the principle of plenitude as being propitious because it maximized diversity.[81] Rolston, coming from a different tradition and unaware of either Leibniz or Lovejoy, similarly highlights the importance of "possibility spaces" and of relationships for "interaction phenomena."[82]

In order to know the good (a *quale*), it must be contrasted with something (another *quale*) to set it apart. In antiquity God was conceived of as an immense mass, distributed throughout the universe. In late antiquity St. Augustine initially adopted this notion of the divine but opposed evil to it as a similarly autonomous substance, albeit entirely inferior to the divine substance. If we cannot conceive of nothing, how can we define evil as nothingness and God as something? Leibniz and Spinoza wondered. A way out of the dilemma was for them to speak in terms of relationship, community, and positionality.

Benedictus de Spinoza (1632–1677) addressed the concept of dynamic tension by aligning the essence of evil with inadequacy, incompleteness, and all that individual beings lack. Obviously, without consciousness we—like Adam and Eve—would not know that we are (metaphorically speaking) naked. Spinoza's clearest formulation of the nature of evil is found in his correspondence with the Dutch merchant Willem van Blyenbergh, written between December 1664 and June 1665. Blyenbergh had initiated the exchange by inquiring after the philosopher's thoughts on the nature of evil. Spinoza developed his notion of evil in counterpoint to the concept of the whole, seeing evil not as an essence but as an attribute of relationships. Thus, Spinoza refused to interpret evil as a matter of the will. Especially important in this regard is his letter of January 5, 1665, in which he defines the evil (*das Böse*) inherent in Adam and Eve's Fall from paradise as the loss of a state of greater perfection.[83]

Obviously, this definition is premised on our notion of what constitutes perfection. It is noteworthy that Spinoza writes *more perfect* (*vollkommeneren*), because the comparative form implies a series of *relative* states of perfection. In fact, we develop a sense of perfection via comparisons and ever more refined differentiations. Ultimately, then, even our idea of perfection is relative. Thus, Spinoza concludes that evil—that is, sin—is merely a sign of imperfection and "does not constitute something that expresses a [concrete] reality."[84] Enhanced consciousness is the path toward unfolding the potential reality. Placed in this light, evil seems to connote a nonexistence or a state of not-yet-realized actuality. Adam's (and Eve's) eating of the forbidden fruit could then be seen, Spinoza suggests, as neither evil nor contrary to the will of God. Partaking of the apple was ultimately an expression of what was yet to come. According to Spinoza, the consciousness of inadequacy appears as such only to the mind of man, not to that of God, to whom all things are perfect and full.[85]

In other ways, Spinoza's argument seems to anticipate the fundamental notion of modern science's holographic universe. To be sure, he ascribes beauty and order, ugliness and chaos, to functions of human perception rather than to operations of nature.[86] Yet while granting that he cannot judge the exact mechanism that unites the parts of nature into a harmonious whole, he thinks (here in agreement with Leibniz) that they result from agonal coordination of conflicting forces "so that they will exist as little as possible in contradiction to one another."[87] To illustrate his point, Spinoza cites the example of the human circulatory system (*das Blut*). If a tiny virus (*Würmchen*) living in the circulatory system were endowed with the powers of sight and discernment, it would be able to observe how the individual corpuscles interact in the turbulent flow coursing through our veins. Living within this closed system without knowledge of any external forces impacting the movement within the veins, the microbe would consider each blood cell encountered to be a whole unto itself, without connection to anything else except its own movement. Each opposing force, each resistant body, would appear always to be a self-sufficient unit. We know, however, that the circulatory system is enclosed in a larger, more complex, interactive system. In just the same way, the part of the universe that humans inhabit and chart is only a part of a larger context.[88] Its modifications (*Veränderungen*), moreover, are limitless.[89] And because the nature of the divine essence is infinite, each of the individual parts is integrated in the wholeness of that one undivided presence. It, in turn, could not be what it is without them.

Then, too, the intellect is part and parcel of nature. The human mind, limitless in its capacity for thought, can bring about change in the physical realm.[90] Consequently, there appears to be no hard and fast distinction between mind and matter for Spinoza. Because of these limitless possibilities of thought and matter, the proper perspective for attaining the level of perfection is the most inclusive perspective possible, not only horizontally—that is, outward in all directions—but also vertically, along the scale from the very small to the very large.

Spinoza's relational explanation of the nature of evil is striking in its modernity. Drawing on the scientific advances of his day and encouraged by his own practical experiences as a lens maker, Spinoza assigned perspective a central role in his practical philosophy. Because of perspectivism he argued for the relativity of such fundamental valuations as good and evil, order and chaos, in an age otherwise known for its dogmatism. Such value pairs were not oppositional by nature; they were actually complementary. In short, Spinoza emphasized very early on "a new feeling for proportion of compatibilities at any given time, of that which is balanced, that which is appropriate for us and everything else."[91] The traditional virtue of modesty gave way to propriety—that is, the hegemony of social decorum was challenged by a new consciousness of ecological appropriateness.

Obviously, Spinoza proposed some heretical ideas in his time.[92] Primary among them was the notion that there is but a single substance with an infinite number of attributes and that all creatures are modifications of this substance—that is, they are modes of its existence. All of these concepts are summed up in the now famous formulation *deus sive natura*. Because of this identification of nature with divine self-regulation, so asserted by Spinoza in the twenty-ninth proposition of the *Ethik*, nature determines the manner in which all things exist and function; there is no room for accident.[93] By grounding ethics in social practice rather than in a transcendent essence, the *Ethik* opens up new territory. Whereas morality represents a judgmental system based on God's pronouncement to Adam and Eve to obey his command, ethics marks a reversal of that judgmental system. Adam and Eve elect to know things on their own—that is, to make their own judgments. It thus signifies a reevaluation of values usually associated with Friedrich Nietzsche.[94]

Key in all this is Spinoza's dualistic view of nature: it has a created and a creative side, a passive and an active dimension. Active nature (*naturende Natur*) reveals "attributes of substance that express an eternal and infinite essence" (*Attribute der Substanz, die ewige und unendliche Wesenheit ausdrücken*).[95] Nature as process is self-regulative, by virtue of which the modi of existence are made possible. Nature as product (*genaturte Natur*) designates for Spinoza, on the other hand, "the total sum of the modes of divine attributes."[96] Attributes, by the way, are those qualities that we comprehend rationally through differentiation.[97] The manner of existence is dependent on the immanence of God.

Because a rational being's intellect links it with the deity, the rational being is simultaneously finite and infinite. However, the more a thinking being is able to use its powers of discernment to recognize more and more of the attributes of God and of nature, the more complete, and, therefore, the more real, it is.[98] Earlier Spinoza had declared that "the more reality or being a thing possesses, the more attributes it will have."[99] In other words, heightened consciousness leads to heightened reality; combined, they would seem to translate into a supracritical state of more-concentrated perfection. The prerogative of the mind, as Leibniz had proffered, is an especially apt version of the divine machine. The end point of Spinoza's philosophy was the intellectual love of God, which he equates in the first part of the *Ethics* with intuition and identifies as the "third kind of knowledge" to distinguish it from sense experience (the first kind) and the reflective knowledge that arises from analysis of sensations (the second kind). Because of this more-differentiated view of the sorts of knowledge, Spinoza seems to retract an exact alignment of nature with God, allowing for the existence of the soul after the decomposition of its material casing. The intuition of something more complete, something better, acts like a homing beacon in a manner analogous to Leibniz's *Monadology*.[100] Thus, God did not err in giving Adam and Eve an imperfect will, for humankind's nature is to seek to know the mind of God, not to actually divine it. Within the overall scheme of things, the human will is thus exactly as it should be. Spinoza underscores his point by drawing an analogy: a circle and a sphere are two different things, somewhat similar in kind, but not self-same.[101] It would be ludicrous to blame the circle for being something it is not. Each rung of enhanced consciousness marks greater perfection. Enhancement of the communication networks represents a fuller realization of the inherent potential in the whole.

In Spinoza's system of ethics no act is *eo ipso* good or bad. Depending on the circumstances, it can be either. Priority is always accorded the active principle in nature and in humankind. Deleuze remarks insightfully that being good is "a matter of dynamics, of capacity, of the constitution of power."[102] Practical philosophy for Spinoza, then, as it seemingly also was for Leibniz, amounted to a way of life, a "*mode d'existence*." The more consciously one is capable of living, the higher the quality of one's existence, and, thus, the more perfect the existence.[103] Life is then a test of the limits of one's capacity for life rather than a preliminary state of existence intended to purify the individual, making her worthy of eternal reward for a life well lived. Life is its own reward, because God chose to instanticize this world as the most apt of all those she conceived of, because it allows for the greatest potential.

Matthew Stuart concludes, in his very readable account of Leibniz, Spinoza, and the fate of God, that the two provocative philosophers "remain unsurpassed today as representatives of humankind's radically divided response to the set of experiences we call *modernity*."[104] Together they supply the basic theory for liberal political order, the underpinnings of modern science, and nonlinear rationality. The reactive form of modern views of the way the world works began with Spinoza and Leibniz, who were motivated to impart meaning to an apparently meaningless universe. Has the search for the seat of consciousness returned us to essential questions raised by this courtier and heretic? David Chalmers aptly remarks in his theory of consciousness, which draws heavily on the role of qualia: "Consciousness is the biggest mystery. It may be the largest outstanding obstacle to our quest for a scientific understanding of the universe." To explain it, we might have to refer to panpsychism—that is, the counterintuitive notion that everything is conscious—and to a quantum-mechanical view of the rise of consciousness.[105] Chalmers does not cite Leibniz or Spinoza.

Notes

1. Andrew Newberg, Eugene D'Aquili, and Vince Rause, *Why God Won't Go Away: Brain Science and the Biology of Belief* (New York: Ballantine Books, 2001), 28–32; Dean H. Hamer, *The God Gene: How Faith Is Hardwired into Our Genes* (New York: Doubleday Books, 2004).

2. See Powers of 10, at http://powersof10.com/index.php?mod=power_detail&id_power (accessed November 14, 2008).

3. See Peter-André Alt, "Kartographie des Denkens. Literatur und Gehirn um 1800," in *Scientia Poetica: Literatur und Naturwissenschaft* (Göttingen: Wallstein, 2004), 163–92, on Gall, 167–73.

4. See Karl R. Gegenfurtner, *Gehirn und Wahrnehmung*, 2nd ed. (Frankfurt: Fischer Taschenbuch Verlag, 2004); Eric R. Kandel, *In Search of Memory: The Emergence of a New Science of Mind* (New York: Norton, 2006); Christof Koch, *The Quest for Consciousness: A Neurobiological Approach* (Engelwood Colo.: Roberts & Co., 2004); Roger Penrose, *Shadows of the Mind: A Search for the Missing Science of Consciousness* (Oxford: Oxford University Press, 1994).

5. K. Brodmann, "Physiologie des Gehirns," *Neue Deutsche Chirurgie* 11 (1914): 85–426. Cf. Koch, *Quest for Consciousness*, who adapts Brodmann's illusrations (118). The map reproduced here is drawn from the University of Michigan Cognitive Science Laboratory website, http://www.umich.edu/~cogneuro/jpg/Brodmann .html (accessed November 14, 2008). For a brief history of brain imaging through the ages, see Edwin Clarke and Kenneth E. Dewhurst, *An Illustrated History of Brain Function: Imaging the Brain from Antiquity to the Present* (Novato, Calif.: Norman Publishing, 1996).

6. Howard N. Tuttle, *Human Life Is Radical Reality: An Idea Developed from the Conceptions of Dilthey, Heidegger, and Ortega y Gasset* (New York: Peter Lang, 2005), 97–100, 153–59; here 158.

7. See my discussion of these issues in John A. McCarthy, *Remapping Reality: Chaos and Creativity in Science and Literature (Goethe-Nietzsche-Grass)* (Amsterdam and New York: Rodopi, 2006), 111–34.

8. Edward O. Wilson, *Consilience: The Unity of Knowledge* (New York: Alfred A. Knopf, 1998); Stuart Kauffman, *At Home in the Universe: The Search for Laws of Self-Organization and Complexity* (New York: Oxford University Press, 1995); Ernst Fuhrmann, *Was die Erde Will. Eine Biosophie* (Munich: Matthes and Seitz Verlag, 1986).

9. Stubenberg, *Consciousness and Qualia*, 126–28.

10. Koch, *Quest for Consciousness*, 15, 232–35, 332.

11. Koch, *Quest for Consciousness*, 22.

12. See the summary at http://www.umich.edu/~cogneuro/jpg/Brodmann.html (accessed November 14, 2008).

13. Koch, *Quest for Consciousness*, 23.

14. Koch, *Quest for Consciousness*, 288.

15. Penrose, *Shadows*, 23, 178–79, 214–16, 357–71. Another more recent physiological approach is Kandel's *In Search of Memory*.

16. Newberg et al., *Why God Won't Go Away*. Koch does not seem to know this work. For a physiological study of the neural networks essential for understanding the biological basis of affect-thinking and feeling, see Larry W. Swanson, "Anatomy of the Soul as Reflected in the Cerebral Hemispheres: Neural Circuits Underlying Voluntary Control of Basic Motivated Behaviors," *Journal of Comparative Neurology* 493, no. 1 (December 2005): 122–31.

17. Koch, *Quest for Consciousness*, 338.

18. Koch, *Quest for Consciousness*, 129.

19. Koch, *Quest for Consciousness*, 123. On hierarchic structure see Koch, 117–31, and the figures of the division of the neocortex according to Korbinian. Cf. http://spot.colorado.edu/~dubin/talks/brodmann/brodmann.html (accessed November 14, 2008) and http://en.wikipedia.org/wiki/List_of_regions_in_the_human_brain (accessed November 14, 2008).

20. Penrose, *Shadows*, 363–64. While Penrose approaches microanatomy in different fashion, he too concludes that the bridge-like connections between cytoskeletons known as microtubule associated proteins (MAPs) ensure an interactive and fully integrated sharing of information also laterally.

21. Newberg et al., *Why God Won't Go Away*, 42–46.

22. Newberg et al., *Why God Won't Go Away*, 45.

23. Koch, *Quest for Consciousness*, 129–30.

24. Newberg et al., *Why God Won't Go Away*, 46.

25. Newberg et al., *Why God Won't Go Away*, 47.

26. Newberg et al., *Why God Won't Go Away*, 46–53; the latter all without specified locality.

27. Newberg et al., *Why God Won't Go Away*, 53.

28. Koch, *Quest for Consciousness*, 8.

29. Koch, *Quest for Consciousness*, 247.

30. Koch, *Quest for Consciousness*, 309–10.

31. Koch, *Quest for Consciousness*, 310.

32. Koch, *Quest for Consciousness*, 310.

33. Leopold Stubenberg, *Consciousness and Qualia* (Amsterdam and Philadelphia: John Benjamins Publishing Co, 1998). See especially chapter 10, "Consciousness: The Having of Qualia," 262–310.

34. Koch, *Quest for Consciousness*, 241.

35. William Barrett, *Death of the Soul: From Descartes to the Computer* (Garden City, N.Y.: Anchor Press / Doubleday, 1986). Barrett focuses on physics and chemistry, lamenting the absence of the mind and of consciousness in those disciplines as he interprets them. His book is actually a plea to reinstate consciousness as a central object of inquiry. He is unaware of neuroscience.

36. Gottfried Wilhelm Leibniz, *Philosophical Writings*, ed. G. H. R. Parkinson, trans. Marry Morris and G. H. R. Parkinson (London: J. M. Dent, 1997), 116. All citations will be from this readily available edition.

37. The *Monadology* is cited by axiom or paragraph number since each entry is brief. This method of citation will allow the reader to locate the source in any edition. Leibniz, *Monadology* §1–3, 7–9.

38. New System, Leibniz, *Philosophical Writings*, 121.

39. Leibniz, "Principles of Nature and Grace," in *Philosophical Writings*, 195.

40. Leibniz, *Monadology* §76–70.

41. Leibniz, *Monadology* §73.

42. Leibniz, *Monadology* §71.

43. Leibniz, *Monadology* §10.

44. On the garden motif see Werner Schneiders, "Gottes Garten. Zu Leibniz' Idee einer Seinsharmonie," in *De Christian Wolff a Louis Lavelle: Metaphysique et Histoire de la Philosophie* (Hildesheim, Zurich, and New York: Georg Olms Verlag, 1995), 3–15. An insightful evaluation of the innovativeness of Leibniz's thought in the context of his times is provided by Ernst Cassirer, *Philosophy of the Enlightenment*, trans. Fritz C. A. Koelln and James P. Pettegrove (Boston: Beacon Press, 1955), 80–92 et passim.

45. Cassirer, *Philosophy of the Enlightenment*, 82–83.

46. Leibniz, *Monadology* §13.

47. Leibniz, *Monadology* §18.

48. Leibniz, *Monadology* §17.

49. Leibniz, *Philosophical Writings* 121; emphasis in original.

50. Leibniz, *Monadology* §14.

51. Leibniz, *Monadology* §29; emphasis in the original.

52. Leibniz moved beyond the Medieval creationism of Alan of Lille, who believed God had to infuse a soul into each child at birth, and also beyond the theory of soul propounded by St. Augustine, who much preferred the soul over the body and was essentially more interested in God than in the connection between the body and the soul. This dualism persisted to Descartes. See Katharina Philopowski and Anne Prior, eds., *Anima and sêle: Darstellungen und Systematisierungen von Seele im Mittelalter* (Berlin: Erich Schmidt Verlag, 2006), xv–xx.

53. Leibniz, *Monadology* §61.

54. Leibniz, *Monadology* §40.

55. Leibniz, *Monadology* §48.

56. Schneiders, "Gottes Garten," 10: "er ist der Höhepunkt dieser Seinssteigerung . . . er ist als Inbegriff aller möglichen und wirklichen Wesensheiten auch deren Ursprung und Urgesetz, also ratio im Sinne von Verhältnis und Grund."

57. Cited by Schneiders, "Gottes Garten," 15.

58. Cited by Schneiders, "Gottes Garten," 7.

59. To be sure, there are those who see, for example, the universe expanding beyond all comprehension because dark matter (which comprises about 90 percent of the universe) is accelerating the expansion so rapidly that "in one hundred billion years the only galaxies left visible in the sky will be the half-dozen or so bound together gravitationally into what is known as the Local Group." Everything else will have slipped over a horizon and beyond human detection. Our own part of the universe will probably, also because of gravitational pull, converge into one starry ball.

Scale obviously plays a role here: the universe is ca. fourteen billion years old, the Earth ca. four million, and the sun is expected to explode in ca. four billion years. Even with an optimal life expectancy, humans are shorter lived than fireflies on that time scale. Nor are we very large. Some see inflation as a sign that we live in a messy universe. Cf. the essay by Dennis Overbye, "The Universe, Expanding Beyond All Understanding," *New York Times*, Tuesday, June 5, 2007, D2.

60. McCarthy, *Remapping Reality*, 109–10, et passim.

61. Penrose, *Shadows*, 216.

62. Penrose, *Shadows*, 357–58.

63. Leibniz, *Monadology* §62.

64. Leibniz, *Monadology* §63.

65. Leibniz, *Monadology* §82, *Philosophical Works*, 192.

66. Leibniz, *Monadology* §79, *Philosophical Works*, 192.

67. Leibniz, "New Essays on the Human Understanding," in *Philosophical Works*, 155.

68. McCarthy, *Remapping Reality*, 62.

69. Leibniz, *Monadology* §29.

70. Leibniz, *Monadology* §30. Cf. also his reflections in "Metaphysical Consequences of the Principle of Reason" (1712), *Philosophical Works*, 174–75.

71. Leibniz, *Monadology* §83.

72. Leibniz, *Monadology* §87.

73. Leibniz, *Monadology* §84, *Philosophical Works*, 193.

74. See Glockner's commentary in Gottfried Wilhelm Leibniz, *Monadologie*, trans. with intro. and commentary Hermann Glockner (Stuttgart: Reclam, 1963), 40–42 and 56–57.

75. Leibniz would have been pleased to learn that modern science emphasizes the importance of relations in determining reality. See, for example, Ilya Prigogine and Isabelle Stengers, *Order out of Chaos: Man's New Dialogue with Nature* (Boulder, Colo.: New Science Library / Random House, 1984); and George Smoot and Keay Davidson, *Wrinkles in Time* (New York: W. Morrow, 1993).

76. Leibniz, "New System," in *Philosophical Writings*, 116–17. While some emphasis is in the original, I have added "force," "detail of the economy of the animal's body," and "explain the particular problems of nature."

77. Leibniz, *Monadologie*, 56: Glockner explains, "Man kann sich die Monaen als *Strahlkräfte* vorstellen, welche von einem immateriellen und schlechtedings einfachen Kern ausgehen und deren Wirkung sich—dem Kontinuitätsprinzip entsprechend—allenthalben durch den ganzen Kosmos erstreckt. Überall im Kosmos ist ein solcher Kern oder Weltmittelpunkt; denn es ist alles lückenlos mit einfachen Substanzen 'angefüllt.'"

78. See Barrett's "maps of the modern world," *Death of the Soul*, 54, 58.

79. Holmes Rolston, *Genes, Genesis and God* (Cambridge: Cambridge University Press, 1999), 52–53.

80. Edward O. Wilson and Stephen R. Kellert, eds., *The Biophilia Hypothesis* (New York: Island Press / Shearwater Books, 1993). See also Wilson, *Biophilia* (Cambridge, Mass.: Harvard University Press, 1984); and Wilson, *Consilience*, 78–81.

81. Arthur Lovejoy, *The Great Chain of Being: A Study of the History of an Idea*, William James lectures originally delivered at Harvard University in 1933 (Cambridge, Mass.: Harvard University Press, 1936), 180–81.

82. Rolston, *Genes*, 350, 354.

83. Benedictus de Spinoza, *Briefwechsel, Sämtliche Werke*, ed. Manfred Walther, trans. Carl Gebhardt, 8 vols. (Hamburg: Felix Meiner Verlag, 1989), 6:81 (letter #19).

84. Spinoza, *Briefwechsel*, 6:80.

85. Spinoza, *Briefwechsel*, 6:82.

86. Spinoza, *Briefwechsel*, 6:146.

87. "daß sie so wenig wie möglich im Gegensatz zueinander stehen"; Spinoza, *Briefwechsel*, 6:146.

88. Spinoza, *Briefwechsel*, 6:147–48.

89. Spinoza, *Briefwechsel*, 6:148.

90. Spinoza, *Briefwechsel*, 6:148.

91. Manon Andreas-Grisebach, *Eine Ethik für die Natur* (Zurich: Ammann Verlag, 1991), 218. In this new ethic of nature, the traditional virtue of modesty required initially by God gives way to the virtue of propriety (*Angemessenheit*) as required by the natural system (219).

92. Because of his radical views, Spinoza was excommunicated in 1656. For an evaluation of his life and times, see Wilhelm Schmidt-Biggemann, *Baruch de Spinoza 1677–1977: Werk und Wirkung*, Ausstellungskatalog der Herzog August Bibliothek, no. 19 (Wolfenbüttel: Felix Meiner, 1977); Giles Deleuze, *Spinoza: Philosophie Pratique*, 2nd expanded ed. (Paris: les Éditions de Minuit, 1981); and Matthew Stewart, *The Courtier and the Heretic: Leibniz, Spinoza, and the Fate of God in the Modern World* (New York & Oxford: Oxford University Press, 2006).

93. Baruch de Spinoza, *Die Ethik nach geometrischer Methode dargestellt*, in *Sämtliche Werke*, 8 vols., trans. Otto Baensch (Hamburg: Felix Meiner Verlag, 1989), 2:31. Hereafter cited in the text by volume and page number.

94. Deleuze, *Spinoza*, 27–43.

95. Spinoza, *Ethik*, 2:32.

96. "die gesamten Modi der Attribute Gottes"; Spinoza, *Ethik*, 2:32.

97. Spinoza, *Ethik*, 2:3.

98. Spinoza, *Ethik*, 2:51.

99. Spinoza, *Ethik*, 2:10.

100. Stewart, *Courtier*, 177.

101. Spinoza, *Ethik*, 6:83.

102. "Car a bonté est affaire de dynamisme, de puissance, et decomposition de puissance"; Deleuze, *Spinoza*, 35.

103. Deleuze, *Spinoza*, 59: "Telle est danc la différence finale de l'homme bon et de l'homme mauvais: l'homme bon, ou fort, est celui qui exise si pleinement ou si intensément qu'il a conquis de son vivant l'éternité, et que la mort, toujours extensive, toujours extérieure, est peu de chose pour lui."

104. Stewart, *Courtier*, 310.

105. David J. Chalmers, *The Conscious Mind: In Search of a Fundamental Theory* (New York and Oxford: Oxford University Press, 1996), xi, 152, 119–21, 333–34.

12

Looking Forward: The Question of Brain, Mind, Self, and Soul

Volney P. Gay

WE HAVE CONFRONTED the conceptual gulfs that separate the sciences from religion and religious experience. Until recently, the terms *brain, mind, self,* and *soul* referred to different entities studied by different methods. Now, for the first time in history, methods gathered under the rubric *neuroscience* seem to aim at a unified goal: the objective study of religious subjectivity.

To bridge the gaps between the processes that scientists study and those that religionists study, we arranged our subject into discrete levels and distinctive disciplines. To articulate these disciplinary levels we brought together two physicists (Christof Koch, S. Victoria Greene), a psychologist (Sohee Park), a philosopher (Alicia Juarrero), a neuroscientist (Jeff Schall), a psychiatrist (Stephan Carlson), an anthropologist (Thomas Gregor), a sociologist (Gary Jensen), a historian (Michael Bess), two religionists (Volney Gay, Edward Slingerland), and a scholar of German and European thought (John McCarthy).

While there are many definitions of *religion*, each turns on the theme of self-consciousness. Even if with Francis Crick (and other atheists) we deny there is a Creator, we know—existentially—that we are created; we did not create ourselves. With that experience come the questions raised by Immanuel Kant, the great German philosopher: What can I know? What ought I to do? For what may I hope? What is a human being? The linking element in these questions is the *I* who asks them.

But what is this *I*? Across human history and human cultures, *this* question has been addressed by narratives. At the center of both small and great religious traditions are stories about our origins and our destiny. How could it be otherwise? We know that our species cannot survive without parenting, and parenting requires that one generation provide an emotional location for the next. Stories about the child, about the child's past, about the child's place in the order of things, are essential to the child's development into a healthy and secure adult. Stories have a beginning, middle, and end. Religion is a grand narrative about our past, our present, and our destiny.

Surely grownups should abandon this childish need for narrations, no? They should recognize what Freud and other critics have said: we must attend to the calm words of science and abandon childhood illusions. Using terms from this book, we should recognize that religious accounts of the soul are illusions. Stories about the soul are really accounts of the self, and those, in turn, are accounts of the mind, the name we give to what the brain does. To condense this even further, we should recognize that when people talk about the soul they are really talking about their brains. To be even simpler, when a person talks about *soul* it is the person's brain shaping verbal behavior about *its* hidden functioning.

When pushed to this extreme—soul reduced to self and mind reduced to brain—the promise of a neuroscience of religion seems fatuous, for the brain is an organ encased within the skull, while persons (or selves) are deeply intertwined with other persons (or selves). The disciplines that make up the neurosciences are as variegated and diverse as the disciplines that make up the university. That they are not fully reducible to one another suggests that our subject matter—the mind—is itself diverse and irreducible. This should not surprise us, for the insight that we are multiple, that we are not unitary beings, is as ancient as human writing.

The Brain and the Self Are Multiple

Both neuroscientists and humanists affirm that for each of us, the brain and the self are divided. The feeling that our self is a unified entity in which conscious reasoning directs the rest of our behavior is an illusion that has been pierced thousands of times for thousands of years. Plato documents the division of human being and then argues that the intellectual part should rule the lesser parts. The goal of philosophy and the good state, he tells us, is to promote that part that seeks justice, "the fair and honorable things being those that subject the brutish part of our nature to that which is human in us, or, rather, it may be, to that which is divine, while the foul and base are the things that enslave the gentle nature to the wild."[1] In the simpler language of the *Phaedrus*, the human being (the human soul) is like a pair of winged steeds; "one of them is noble and good, and of good stock, while the other has the opposite character."[2]

Contemporary neuroscientists use modern metaphors to make the same point. Rather than speak of winged steeds, they speak about parallel computational devices or multiple brain sites. Robert Prentky lists more than forty authors who have described two ways of knowing or "two cognitive styles." He cites Buddhist theologians, philosophers, Freud, anthropologists, and neuroscientists who study human cognition.[3]

That we know this and yet do not always live it illustrates the same phenomenon: we are driven to presume a unity to the self even though science tells us it is an illusion. Because the self is multiple, there must be boundaries within it. Neuroscience tells us that the systems responsible for emotional arousal are separate and independent of the higher systems—such as conscious, deliberate thought. Parts of the self (of the brain) are unconscious, nonverbal, and organized in modules that do not link directly to conscious, reasoning modules. This suggests limits to self-understanding and to empathy.

To be empathic is to comprehend motives and other mental events causally related to a person's choices: whatever *drives* a person to act cannot be comprehended empathically. *Internal world* and *external world* name parts of my environment—the surround—that lay beyond my motivation. To believe that we can empathically comprehend those internal and external worlds is a form of animism. We recall nineteenth-century scholars of religion who examined parallel beliefs in diverse religions, such as E. B. Tyler, J. G. Frazier, and Wilhelm Wundt. They noted that animists view the world as operating on motivational systems analogous to those found in persons. If everything around me contains spirit or soul, then everything is subject to wishes, desires, and supplication. Animism is seductive because it helps us maintain this grand illusion. Animism supposes not only that nature—the "out there"—is subject to motivations, wishes, and meanings, but that the "in here"—the interior parts of the self or brain—is equally controlled by motivations and wishes. We overpersonify persons by ascribing to their actions, motives, and wishes that spring from diverse parts of the brain.

In their own ways, religious authorities recognize the multiplicity of the self, and they use paradoxical formulations to describe it. For example, when Jesus and the gospel authors speak about the end of time, they refer to two forms of time: ordinary time, measured by bodily experiences, and extraordinary time, time beyond the chronological sequence. This second time will occur or has occurred already; it all turns on how we conceive of ourselves in-depth—that is, our souls and their location in the kingdom. In the "Jesus Seminar," all the sayings attributed to Jesus were scrutinized, and few were chosen.[4] Among those that made the grade we find the parable of the mustard seed and the parable of leaven.[5] In both, Jesus uses a homey metaphor to say something about the kingdom of God: it is like a mustard seed and also like leaven, for both are small, apparently insignificant, entities that grow large—the first a refuge for birds, the second a transformative power. Both of these metaphors suggest that Jesus believed that he heralded a new form of time and was ushering in God's imperial rule.

Reflecting on these and other eschatological sayings, the Apostle Paul says that while Jesus's sayings may seem obscure now, in the final days we will see everything clearly: "Now I know in part; then I shall understand fully, even as I have been fully understood."[6] These famous lines promise eventual understanding, not current lucidity. We will understand what the "end of time" means at the End of Time and not before. Before that event we must remain content with the promise of eventual resolution.

To remain faithful to their creed, atheist philosophers must dismiss Paul's metaphors as signs of intellectual confusion, or deceit, or error. To remain faithful to their creed, Christian theologians wager that Paul's intellectual struggles merit consideration. In making that wager they extend Paul's symbolic utterances without collapsing them into mere folly. For example, Rudolph Bultmann discusses Paul's disquisition about Jesus and the end of time and acknowledges that Paul's language is a "mirrored image full of riddles."[7] Paul cannot escape contradiction when he attempts to explain the doctrine of resurrection. Sometimes Paul says that the resurrection of saints will occur at the end of time, that moment in human history when this world as we know it comes to an end. At other times Paul seems to say that resurrection will occur upon one's death. Paul affirms both that the resurrection of all believers is in the near future, perhaps in his generation, and that Christians are already resurrected and living in the spirit of Christ. When added together, these two propositions form a contradiction: "It is no longer I who live, but Christ who lives in me."[8]

Brain, Self, Soul

In opposition to the atheists' wholesale dismissal of religion, modernist theologians like Bultmann seek to demythologize ancient biblical texts and to recover the essence of Jesus's teachings. As Bultmann says, to demythologize is "not to eliminate the mythological statements but to interpret them."[9] To do this, Bultmann reinterprets Jesus's message that the "kingdom is at hand" along with similar pronouncements that are eschatological and, superficially, false. For history did not culminate in the death of Jesus. Relying on the philosophical anthropology of his contemporaries, especially Martin Heidegger, Bultmann offers what he claims is a nonmythological reading of Jesus's proclamation. In *History and Eschatology: The Presence of Eternity*, Bultmann sums up: "It is the paradox of Christian being that the believer is taken out of the world and exists, so to speak, as unworldly and that at the same time he remains within the world, within his historicity."[10]

Does Bultmann's reading escape the ancient myth and the circle of metaphorical reasoning? Yes and no. Yes, like other modernists Bultmann rejects the näive worldview of the biblical period and therefore does not affirm biblical literalism like that found in Creationist theology. No, when Bultmann uses contemporary philosophy to discern a meaning of Jesus's parables, for example, he cannot escape their paradoxical gravity. Bultmann takes up the red thread that runs through both personal and public mythology, the problem of time and our struggle to comprehend how in the future we might be other than we are in the present. Religious fundamentalism grasps one part of the paradox, that human beings are multiple and therefore incomplete, but rejects the other, that we cannot know our destiny. Bultmann reflects on the errors of biblical literalists who claim that "the content of self-understanding is a timeless truth; once perceived it remains valid without regard to the occasion, namely, revelation, which has given rise to it."[11]

Bultmann challenges fundamentalists who claim to have isolated a single, nonparadoxical meaning to Jesus's admonitions. All versions of fundamentalism tend toward hubris and reductionism. From Hegel to Marx to Nietzsche we find consistently the claim that each thinker has summed up the past and has finally "seen through" history. Because the self (or brain, or ego, or person) is diverse, divided, and multiple no nonparadoxical formulation can name it fully.

If we accept Bultmann's interpretation of Jesus's teachings it follows that no method, no matter how seductively garbed as science, will let us predict and control fully who we shall become. This theological and psychological claim is a kind of limiting axiom, a metaphysic grounded on humility. It suggests that we can be either näive fundamentalists (biblical or neuroscientific) who purport to know absolutely or we can recognize our status as creatures captured in existential dilemmas.

The Plasticity of Memory and History

New theories of mind become new ideologies that recast our view of the past and the future.

Adopting Darwinian, Freudian, or neuroscientism requires that one reassess the past and thus fabricate a new version of history and with it new memories. How we conceive of our individual and collective histories shapes how we act now, and that, in turn, shapes our future. For that reason, establishing a "true history" of a person or a nation is a political act. The past is not secure; our version of history, of who we were, is vulnerable to other, competing, interpretations. More so, in line with both neuroscience and ancient wisdom, we must often discover who we are by observing ourselves act over time and under duress.

For example, in "Eli, the Fanatic," a bejeweled story by Philip Roth, Eli Peck, a young, assimilated American Jew, confronts Leo Tzurf, a Jewish refugee who wishes to open a yeshiva for eighteen Jewish children in a comfortable, WASPish suburb that forbids boarding schools.[12] In the story's beginning, Eli angrily pushes Tzurf to obey the city's zoning laws; by the tale's middle, Eli starts to doubt himself; and in the end he finds himself merging with the despised other, especially Tzurf's colleague, a survivor of the camps who has one possession, an ancient black suit that infuriates the town's sense of propriety. Tzurf explains why the black suit matters: "But I tell you, he has nothing. *Nothing.* You have that word in English? *Nicht? Gornisht?*" His family is dead; he has lost his country and his language. Tzurf asks Eli if he can empathize with such loss: "A synagogue where you knew the feel of very seat under your pants? Where with your eyes you could smell the cloth of the Torah?"[13] These beautiful, empathic questions spur Eli into action, and through his actions he discovers something hidden about himself.

Every religion conveys similar truths: We are more than our brain and more than we know. We are more than solitary animals roaming the world. Eli comes to know more about his self (or about his soul, as religious people would put it) by failing to understand his actions. We are drawn into the story by Roth's empathy for Eli, for the outcast Jews, and for the comfortable townsfolk uneasy about the strange yeshiva. We identify with all of the characters—victims and victimizers alike. The story captures us because we share the emotional and cognitive tensions induced in the characters. Like them, we ask, How did this begin? How will it end? We need to know, but we cannot satisfy that need without engaging with the story, just as Eli cannot know himself without engaging with Tzurf and in that struggle watch himself emerge.

Notes

1. Plato, *Republic*, trans. Paul Shorey, in *The Collected Dialogues of Plato*, ed. E. Hamilton and H. Cairns (Princeton, N.J.: Princeton University Press, 1961), 9, 589, d.

2. Plato, *Phaedrus*, trans. R. Hackforth, in *Collected Dialogues*, 246, b.

3. Robert A. Prentky, *Creativity and Psychopathology: A Neurocognitive Perspective* (New York: Praeger, 1980). Prentky cites Blaise Pascal (1623–1662), who spoke about two "kinds of intellect" (64).

4. Robert Funk, ed. *The Five Gospels: The Search for the Authentic Words of Jesus* (Santa Rosa, Calif.: Polebridge Press, 1993).

5. Luke 13:18–19, 20–21.

6. I Cor. 13:12, Rudolph Bultmann, trans.

7. Rudolph Bultmann, *Theology of the New Testament*, 2 vols. (New York: Scribners, 1951–1955), 346.

8. Gal. 2:20, Bultmann trans.

9. Rudolph Bultmann, *Jesus Christ and Mythology* (New York: Scribners, 1958), 18.

10. Rudolf Bultmann, *History and Eschatology: The Presence of Eternity* (New York: Harper, 1957), 152.

11. Bultmann, *Jesus Christ*, 73.

12. Philip Roth, *Goodbye, Columbus and Five Short Stories* (New York: Modern Library, 1959), 249–98.

13. Roth, *Goodbye*, 264.

Index

About the Contributors and Interviewees

Michael Bess is the Chancellor's Professor of History at Vanderbilt University. His books include *Choices Under Fire: Moral Dimensions of World War II* (2006) and *The Light-Green Society: Ecology and Technological Modernity in France, 1960–2000* (2003).

Stephan Carlson is a fellow in psychiatry and law at Columbia University. He earned his B.A. from Johns Hopkins University and his M.D. from Louisiana University. He did his internship and second postgraduate year in internal medicine at N.Y.U. Downtown Hospital, followed by training in psychiatry at Vanderbilt University Medical Center.

Volney P. Gay, editor, is professor and chair of religious studies, director of the Center for the Study of Religion and Culture, professor of psychiatry, and professor of anthropology at Vanderbilt University. He is also a faculty member of the St. Louis Psychoanalytic Institute.

Victoria Greene is professor of physics at Vanderbilt University. She earned her B.A. from the University of Tennessee in physics and math and an M.S., M.Phil., and Ph.D. in physics all from Yale University. Since fall 2008, she has been executive dean of the College of Arts and Science at Vanderbilt. She is the author or coauthor of more than 175 scientific papers.

Thomas A. Gregor is professor and former chair of anthropology at Vanderbilt University. He has made films about the Mehinaku people of Brazil for the BBC, Grenada Television, and NET. He is author of *Mehinaku: The Drama of Daily Life in a Brazilian Indian Village* and *Anxious Pleasures: The Sexual Lives of an Amazonian People.* His edited books include *A Natural History of Peace, The Anthropology of Peace and Nonviolence,* and *Gender in Amazonia and Melanesia: An Exploration of the Comparative Method.*

Gary Jensen is professor of sociology and religious studies at Vanderbilt University. His most recent book is *The Path of the Devil: Early Modern Witch Hunts* (2007). With Ronald Akers, he has explored macro-micro transitions in criminological theory ("Micro-macro Transitions in Criminological Theory: Taking Social Learning Global" in *Social Learning and the Explanation of Crime: New Directions for a New Century,* 2007) and fractal scales in test of historical theories.

Alicia Juarrero is professor emeritus at Prince George's Community College in Largo, Maryland. She received her B.A., M.A., and Ph.D. in philosophy from the University of Miami (Florida). Dr. Juarrero is author of the internationally acclaimed *Dynamics in Action: Intentional Behavior as a Complex System* (1999).

Christof Koch was born in the American Midwest and grew up in Holland, Germany, Canada, and Morocco as a devout Catholic. He studied physics and philosophy at the University of Tübingen in Germany where he earned his Ph.D. in biophysics in 1982. After four years at MIT, he became the Lois and Victor Troendle Professor of Cognitive and Behavioral Biology at the California Institute of Technology. Among his books is *The Quest for Consciousness.* For additional information, see http://www.klab.caltech.edu/~koch.

John A. McCarthy is director of the Max Kade Center for European and German Studies, professor of European studies, and professor of German and comparative literature at Vanderbilt University. McCarthy is the author or editor of thirteen books, eighty-seven articles and book chapters, and forty book reviews. His most recent publications include *Remapping Reality: On Chaos and Creativity in Science and Literature* (2006) and *The Many Faces of Germany: Transformations in the Study of German Culture and History* (ed., 2004).

Sohee Park is professor of psychology and psychiatry at Vanderbilt University. Born in Seoul, Korea, she trained at the University of Cambridge, and earned her Ph.D. at Harvard University, studying the frontal cortex in schizophrenia under the guidance of Philip Holzman. Among her publications are: "Food Preference and Hedonic Judgment in Schizophrenia," in *Psychiatry Research* (at press), and "Social Cognition in Schizophrenia: NIMH Consensus Meeting on Definitions, Assessment, and Research Opportunities," in *Schizophrenia Bulletin* (2008).

Jeffery D. Schall is E. Bronson Ingram Professor of Neuroscience and director of the Center for Integrative and Cognitive Neuroscience (CICN) at Vanderbilt University. He investigates visual and cognitive influences on eye movement. Funding for his research has come from the National Eye Institute, the National Institute of Mental Health, and the Air Force Office of Scientific Research.

Edward Slingerland is associate professor of Asian studies at the University of British Colombia and holds the Canada Research Chair in Chinese Thought and Embodied Cognition. His research includes Chinese thought, cognitive linguistics and metaphor theory, evolutionary psychology, comparative religion and philosophy, and the humanities and science.